REINVENTING CRAFT IN CHINA

First published in 2017 by
Sean Kingston Publishing
www.seankingston.co.uk
Canon Pyon

British Library Cataloguing in Publication Data
A catalogue record for this book is available from the British Library.

Printed by Lightning Source

ISBN 978-1-907774-98-0

Reinventing craft in China

The contemporary politics of Yixing *zisha* ceramics

Geoffrey Gowlland

Publishing
Sean Kingston Publishing
www.seankingston.co.uk
Canon Pyon

Contents

Acknowledgements

This book is based on ethnographic field research conducted in China as part of my doctoral studies in 2003/4, and on return visits in 2006, 2008 and 2012. I wish to express my heartfelt thanks to the artisans of Dingshu who enthusiastically shared their thoughts and stories with me, who invited me into their workshops and homes, and who were helpful and patient in answering the questions of a foreigner who knew so little when he first arrived. I owe special thanks to Zhao Jianghua ('once a teacher, always a teacher'), Ge Taozhong, Pan Chiping and Wang Yinxian for their particular generosity with their time and thoughts. My thanks also to those who got my research off to such a good start: Lu Liyun, Shi Juntang and above all Yu Jinyao, who went out of his way to help, including in difficult moments. Wu Fei provided invaluable help at first in e-mail exchanges, and later as an ideal research assistant during my return visit in 2012. I want to extend my very special thanks to Wu Yaokun, aka Robin, for being a true friend at a time when the experience of ethnographic fieldwork was getting difficult, and for keeping up our friendship despite the few opportunities to meet in person.

I owe much to Alan Macfarlane, my thesis supervisor at the Department of Social Anthropology, University of Cambridge, and his wife Sarah. They both provided invaluable advice and support. Alan taught me so much about the academic career, writing creatively, and encouraged me to pursue my own path, advice that I continue to cherish years after the PhD.

Marilyn Strathern and Charles Stafford acted as my thesis examiners. I have used their immensely useful comments as a basis for transforming my thesis into this book. Charles continued to act as my mentor during a one-year postdoc at the London School of Economics, and provided support and advice in my transition from student to scholar.

I am grateful to the following institutions for financial support of the fieldwork and writing-up activities that resulted in this book: the Royal Anthropological Institute (Sutasoma Award), the Faculty of Archaeology and Anthropology, University of Cambridge (Anthony Wilkin Fund grant), the Department of Social Anthropology, University of Cambridge (grants from the Ling Roth Fund, the Richards Fund and the Rivers Video Project), the UK Economic and Social Research Council (a Postdoctoral Fellowship based at the London School of Economics) and the Research Council of Norway (Yggdrasil research scholarship, for a research stay based at the Cultural History Museum, University of Oslo).

Chapter 3 contains revised material from the article 'Learning to see value: exchange and the politics of vision in a Chinese craft' published in 2009 in

Ethnos 74(2):229–50, reproduced with permission. Chapter 4 is a revised and updated version of the article 'Learning craft skills in China: apprenticeship and social capital in an artisan community of practice', published in 2012 in *Anthropology and Education Quarterly* 43(4):358–71, reproduced with permission. Chapters 5 and 6 contain revised sections of the article 'Style, skill and modernity in the Zisha pottery of China' published in 2009 in the *Journal of Modern Craft* 2(2):129–42, reproduced with permission.

I owe so much to my parents, who provided support and encouragement throughout my studies and research. My wife Marzia shared moments of discovery, enthusiasm, elation, confusion and frustration, and through our discussions has helped me to untangle ideas and make sense of my experiences. She has been and remains my travel companion, in the broadest meaning of the phrase.

Notes on transliteration and names

Interactions and interviews with artisans in Dingshu were mostly conducted in Mandarin Chinese. People in the area speak a Wu dialect, though tend to be fluent in Mandarin, the national language. In the text, I use the standard Pinyin transliteration convention for words in Mandarin Chinese, and occasionally refer to Wu dialect terms rendered into Mandarin Pinyin, or use a transliteration that is not standard Pinyin but resembles more closely the local pronunciation.

I have used both real names and pseudonyms in the text. Artisans are public figures, and their work and lives are extensively talked about on the internet and in specialized publications. I have used real names in cases where what I talk about is either in the public domain or it is expected that I use people's real names. Using real names is also for me a way to acknowledge and publicize the work of talented artisans. At other times in this book I use pseudonyms in cases where artisans spoke to me confidentially, when I refer to observations of interactions or more simply in contexts in which interactions did not take the form of formal interviews. In some cases I have used a pseudonym and a real name for the same person in different parts of the text. When using real names, I first spell out the name in the usual Chinese form – last name (made up of a single syllable) followed by given name (either one or two syllables), and when I repeat the name in the following sentences I use the last name only. For instance, Gu Jingzhou becomes Gu, Mr Gu or master Gu. When using pseudonyms, I precede the name with an English title, for instance Mr Ma, and follow by using only the last name, Ma.

Books on Yixing *zisha* pottery tend to include a list of names of the more famous and noteworthy artisans and a description of their work. I do not take this approach: artisans and masters mentioned in this book are not selected according to the value that I place on their work, and any omissions are not motivated by any judgment on my part. Rather, the people that I mention in the text are persons who came to be acquaintances, informants or friends in the context of my ethnographic fieldwork. My aim in these pages is to give a glimpse of the lives of Chinese artisans, rather than to evaluate their work.

CHAPTER 1

Introduction

Driving from Dingshan to Shushan, what used to be two separate townships in the Jiangsu province of China and now united as Dingshu township, one goes past a long whitewashed wall, in the middle of which stands an imposing gate in classical Chinese style (Figure 1). One might expect such a wall to circle a temple, mansion or government building; instead it is surrounding a rather common-looking 60 metre high hill known as Huanglongshan ('Mount Yellow Dragon'). Huanglongshan is however not just any hill: this is where one of the most prized clays of China is mined. This clay, known as *zisha*, or 'purple sand', named after its dark brown colour with hints of purple, is most famously used to produce *zisha* or Yixing teapots (the latter named after the neighbouring city) (Figure 2), which, depending on their quality and fame of the potter, can often sell for the equivalent of thousands or tens of thousands of dollars. *Zisha* teapots have been valued for centuries as vessels particularly suited to brewing fine tea, and historically they have been strongly associated with China's class of scholars, who were not only users of these teapots but often also patrons of the craft. It is usually claimed that the particular physical properties of the clay left unglazed – its microporosity after firing – is key to explaining its suitability for preparing tea: when brewing, the tea leaves are said to 'breathe' in the teapot. But it is also the skills and knowledge of artisans that are key to producing vessels particularly suitable as teapots. Over the centuries, artisans have developed unique techniques to use a clay unsuited for being shaped on the potter's wheel (due to its high sand content): the leather-hard clay is compressed into sheets and discs by beating it with a wooden flat-faced mallet. The resulting thin walls of the pot, usually between 2 and 3

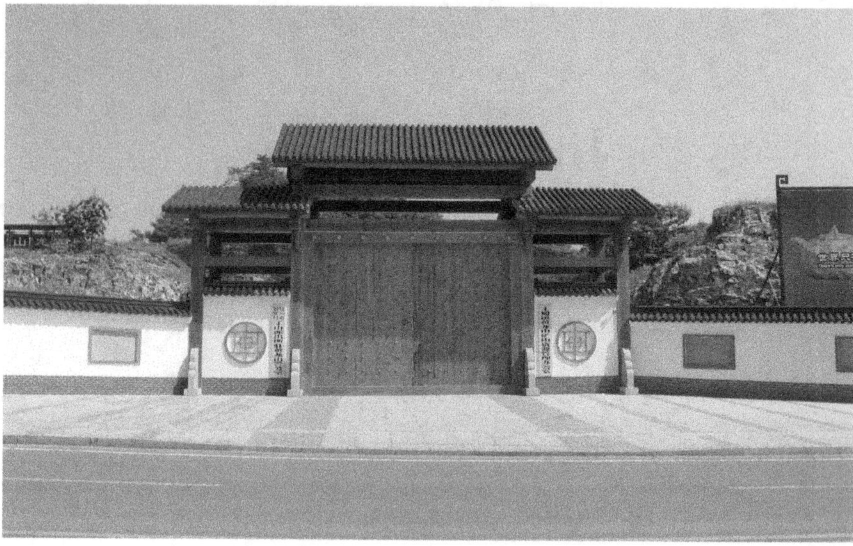

Figure 1 Wall and gate in front of Huanglongshan (2012).

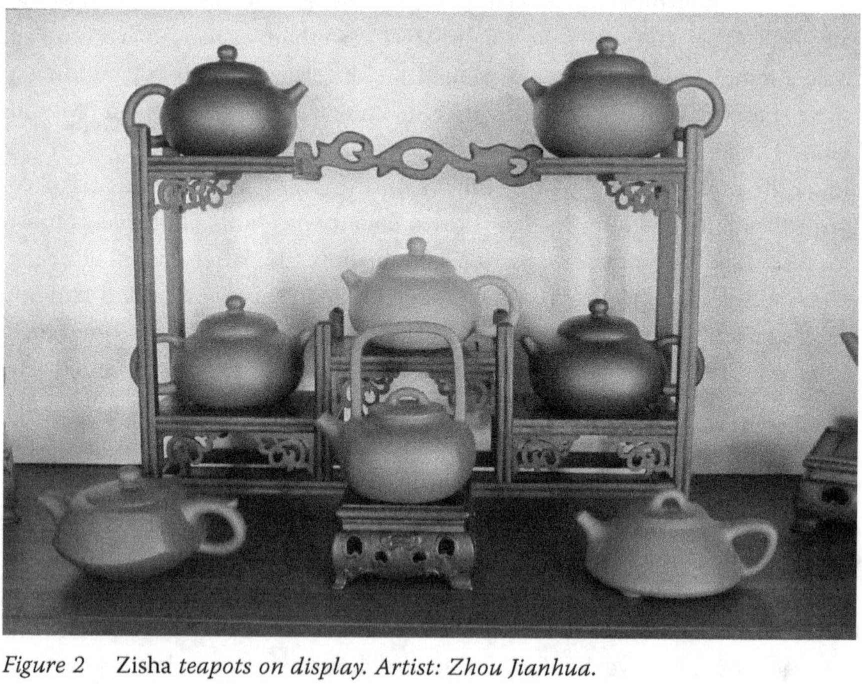

Figure 2 Zisha *teapots on display. Artist: Zhou Jianhua.*

millimetres once fired, further enhances the effect of the microporosity of the clay when tea is brewed.

The value of individual pots is largely related to the fame of the maker, and in this book I will explore issues related to the achievement and construction of mastery and fame and issues relating to authenticity, specifically in relation to a tumultuous recent history. But it is opportune to start this book about *zisha* pottery with the very material that it is made of. Indeed, at the root of the value of these pots is the relative scarcity of *zisha* clay, claimed to be unique to the region, with the best quality found only under Huanglongshan (two other types of clay commonly associated with *zisha* ceramics are also found in the area). At first glance, one might interpret the wall surrounding Huanglongshan as a form of protection of a scarce resource: the wall, erected relatively recently by the local government is intended to enforce the recent decision that exploitation of the mines can only be carried out by government-approved firms, the aim being to help prevent the illegal exploitation of the mines by individual entrepreneurs. Yet this measure does not appear to be particularly efficient: I was told that enterprising individuals can easily get over the wall and dig for clay during the night. Rather than an efficient deterrent, the wall might instead be regarded as a strong claim of the state to a material whose value (monetary, but also symbolic, in relation to a revalorization of China's heritage) has grown exponentially over the past few decades.

At the origin of the construction of the wall is an incident that, in 2010, has highlighted issues related to the creation and maintenance of the value of the clay. In May of that year, the public national television network CCTV aired a news report denouncing the practices of a national electronics company, Meidi, in relation to an electric slow cooker marketed as including a pot made of 'pure *zisha* clay' (see Baidu Baike n.d.a; Cheng 2010). The report accused the company of making false claims: not only was the clay not what is commonly referred to as *zisha* – that is, clay from the area of Dingshu, nor even clay with similar properties – but it was produced by mixing various common clays with chemical additives that gave it the appearance – colour and sheen – of *zisha* clay. Consumers were being tricked into believing that a cheap ceramic pot had been made with valuable clay, and to make things worse, those chemical additives were claimed by the news report to be hazardous to people's health (a statement later put in doubt). The report had immediate consequences not only for the Meidi company (it soon issued an admission of guilt and apologies to consumers), but to the tens of thousands of makers of *zisha* pots in and around the town of Dingshu, due to suspicions that the clay they were using was of the same kind as the Meidi pots, thus carrying the same health risks.

In the immediate aftermath of the report, the marketplaces of Dingshu became deserted. The county government responded with a list of measures in an immediate reaction to the situation, in an attempt to restore the trust of consumers in one of its key products. Besides a crackdown on producers of 'fake' *zisha* clay, it introduced mandatory (and temporarily centralized) chemical analysis of all *zisha* products put on sale. For a period of time, it was also possible for owners of *zisha* ceramics to request an analysis of the clay to determine whether the pots they made contained unhealthy compounds. Artisans themselves soon adopted the habit of analysing the clay they used, and making available a copy of the results to potential customers.

The tightening of governmental control of *zisha* products also led to control of the craft's resources: the act of encircling Huanglongshan with a wall, and making private exploitation of the mines illegal, made government-controlled enterprise the single trusted source for 'authentic' *zisha*. As a result, the cost of *zisha* clay immediately rose, and depending on its quality it reached prices of up to 160 yuan per kilo (about $24), several times the asking price of private dealers before the incident.[1] One artisan even recalled the time when, before the mid 1990s, he could pay the same price for an entire truckload of the same clay.

The news report could have been a catastrophe for the industry, but in the end it does not appear to have had long-lasting economic impact. In 2012, during my latest visits to Dingshu, artisans tended to dismiss the affair as having had little impact on their work, though some pointed out that some factories and poorer producers in rural areas had suffered disproportionately. Yet the event marked significant changes in relations of power between individual entrepreneurs and the state in relation to the control of the raw materials of the craft.

This was not the first time that the state had claimed control of clay resources, though in 2010 it was doing so in a very different context. Soon after Mao Zedong's push to collectivize industries in the 1950s, an enterprise devoted to the extraction and processing of clay became the first state-owned enterprise to be established in Dingshu. Control of *zisha* clay was, however, relinquished in the mid 1990s – a time when China was abandoning many of its state-owned enterprises – when extraction of the clay was not deemed profitable enough and was left to private industry. The construction of the wall around Huanglongshan can be seen as a reclaiming of state control over the raw materials of an industry that has not only thrived and become an important economic resource for the town in recent years, but that also has strong symbolic resonance: in 2008, *zisha* pottery and its methods of production were included in China's national list of protected intangible heritage.

1 All dollar amounts are in US dollars. The exchange rate at the time of my last period of fieldwork in 2012 was about 6.5 Yuan to 1 US dollar.

How does a craft reinvent itself as 'traditional' following times of heightened cultural, social and political transformation? In juxtaposing the terms 'reinvention' and 'tradition', my intention is to capture the two apparently contradictory movements of creating something new whilst inscribing it in a line of continuity with the past. In this ethnographic study, I explore the impact of political, social and economic change in China, and in particular the two modernization projects of the communist regime – collectivization and industrialization in the 1950s, and privatization since the 1980s – on the lives of artisans and processes of transmission of craft-related knowledge. Beyond the regional perspective, the broader aim of the book is, through a single case study, to understand how the specific demands of craft and craft practice maintain a particular kind of relationship with social change in the contemporary world. In the following chapters, my focus will be on the stories of former workers producing ceramics in a factory, and their efforts at reinventing themselves as private producers and artisans, and promoting their work as belonging to an age-old tradition. In the 1950s, a state-controlled factory regrouped artisans producing *zisha* pottery with a view to transform the craft into an industry. Four decades later, starting in the mid 1990s, artisans once more were able to set up private workshops and teach their children, with the consequence that the *zisha* ceramics industry was transformed back into a craft, notably in terms of ownership of the means of production and access to knowledge. I explore the way knowledge that had been made public during collectivization is subsequently transmitted, shared, appropriated and made exclusive in the craft community. One might speak of a process of re-appropriation of the craft by artisans – the techniques and skills associated with the craft, as well as legitimacy in producing and talking about their work. Yet as we have seen with the case of control of *zisha* clay, the state has in many ways retained its presence in craft production, and even extended its influence over recent years. Notably, and despite the privatization of the state-regulated factory and legitimacy of private enterprise, China's government is still involved in the 'art world' (Becker 1982) of *zisha* pottery: with its inscription on the national list of endangered intangible cultural heritage, this unique form of ceramics is caught up in the politics of cultural nationalism in China, and in discourses on the longevity and grandeur of Chinese civilization.

The ethnographic perspective I adopt in these chapters mean I start with a consideration of ways of relating in a craft community – relations between peers, classmates, apprentice and master, urban and rural artisans, and between producers and the state – and a reflection on the way these create the conditions for producing certain kinds of objects, which can be said to have 'agency' (Gell 1998) as they circulate in both material and ideal form and transform those same relations.

'Craft' and crafts in China

Throughout this book, I will refer to the state as a key player in the production of *zisha* craft, be it through the control of the work of artisans or control over raw materials. Handicrafts were particularly politicized in China throughout Mao's government from 1949 to 1976. Within a few years of the communists taking power in 1949, virtually all urban artisans and small producers nationwide were regrouped into collectively owned (but state-controlled) or state-owned factories (see Walder 1984). Through these factories, the state soon had full control over the recruitment of workers, as well as the distribution of raw materials and sales. In the socialist imagination, handicrafts were an anomaly in a modern society, and the Maoist government had clearly spelled-out ambitions to transform the country into an industrialized one where crafts would eventually be fully replaced with mechanized methods of production (Schran 1964). Production followed the guidelines and policies of the state, economic or ideological.

The state's shifting policies regarding export had an obvious impact on production and the sale of products, but so did policies that were not overtly economic. For instance, the campaign to 'destroy the Four Olds' during the Cultural Revolution (1966–1976) aimed at eradicating all traces of 'feudal' society. In the case of *zisha*, this had implications for the kind of products that could be made: products that did not look traditional, and that bore calligraphic inscriptions that were obviously in line with Mao's ideas. State intervention in the crafts of the country was not new at the time. The most striking example of state control of crafts during the Imperial (pre-1911) era was the control by government officials of the porcelain kilns of Jingdezhen, whose products had in significant part been destined for the Imperial court since the Yuan dynasty (1271–1368) (Huang 2008:1). What is particularly striking with socialism is the capillary efficiency with which the government, at local, provincial and state levels, intervened to transform craft production nationwide. As Walder (1984) notes, it took the government a mere eight years to achieve virtually full control over the recruitment of workers, production and sale of industrial products, the majority of industries resulting from the regrouping of small-scale producers into cooperatives and factories. In the frenzy towards the industrialization of the nation at all costs (including that of human lives, such as those lost during the tragic Great Leap Forward, which resulted in famines), handicraft producers were to be transformed into factory workers, unless they resided in rural areas in which case they had to change their activity and become farmers (see Eyferth 2003, 2009).

The politicization of craft production is perhaps less obvious in the context of the turn to capitalism beginning at the end of the 1970s, and which culminated in the privatization of many of the cooperative and state-owned factories during the mid 1990s. As I will discuss throughout this book, however, state presence nowadays is no less real, for instance manifesting itself in policies relating to the protection of heritage. The passage to capitalism marks a transformation, for at least some crafts (those deemed valuable in the context of the rediscovery of heritage), from *shougongye* into *shougongyi*: both these terms are translated as 'handicrafts', but the first alludes to crafts as 'industry' (*ye*) and the second to 'arts and crafts' (*yi*). The first, under Mao, was considered an anomaly, to be upgraded to full *gongye* (industry) status. The second, promoted in the context of a turn to the re-evaluation of heritage, is seen as desirable, not only because it corresponds to the new politics of the Chinese nation-state, which promotes the heritage of the country, but also because of its economic value.

The current renewed emphasis on *shougongyi* is taking place in the context of a general re-evaluation of the past of the country, mingling with a cultural-nationalist discourse, which has an impact on the way certain crafts are viewed. The frenzy in the creation of museums that has characterized the last decade, for instance, attests to a renewed interest in ancient artefacts, especially of the Imperial era (Denton 2005; Varutti 2014). The reasons for such an interest are to be found in disillusionment following the weakening of communist ideals culminating in the dismantlement of communist regimes abroad, coupled with an ideological and identity crisis. In the current context, the ancient past appears as a 'golden age', an era of stability and prosperity, of which cultural relics are the tangible evidence (Varutti 2014). This revival is heavily supported by the government, for which a celebration of the traditions supposedly at the core of Chinese civilization allows for a narrative of continuity aimed at legitimizing its rule (Suresh 2002). The re-evaluation of heritage is therefore no less 'political' in China than the policies of collectivization that aimed to transform artisans into workers. Crafts were most obviously political under Mao, but they remain political in the contemporary context too.

In Dingshu, the ambiguity of craft as both industry and heritage is present in the discourse of artisans, who might talk of the 'industry' (*hangye*) of *zisha*, but also of their work as *shougong*, 'hand work'. It seems that it is important for artisans to recognize *zisha* as an industry, especially after having helped to develop it in the context of a cooperative factory, but it is also important for them to identify the technical process that uses the work of the hands (*shougong*) and not of machines in order to stress the heritage value of the craft. One of the potentials of a study of crafts in China is that it allows us to understand how these changing ideas about craft have played out, and how the status of the artisan has been transformed as part of changes in the evaluation of objects.

This shifting evaluation of crafts resembles the European path: the inefficient methods of craft manufacture were expected to be fully replaced by mechanized production during the industrial revolution. Since the Arts and Crafts movement of the nineteenth century, and seemingly increasingly in contemporary societies, there have been sustained political attempts at protecting, promoting and valorizing craft production. Perhaps the main difference between China and Europe is the speed at which these shifts in the meaning of crafts have taken place. In Dingshu, as elsewhere in China, some of the same individuals who had been identified as 'backwards' handicraft producers in the 1950s and brought to join cooperatives with a view to transforming them into workers, were later, since 1979, encouraged to create works using non-mechanized methods that bore a relationship to the crafting traditions of zisha and to national culture. What is particularly intriguing to me in a Chinese case study on craft is that, although the craft objects remain virtually the same over time, and are created using the same techniques and the same skills, the subjectivities of artisans have been transformed.[2]

Dingshu and *zisha* pottery

Together with Guangzhou, Jiangsu has long been one of the richest provinces of China, with the most developed handicraft and cash-crop economies, as well as extensive involvement in international trade (Faure 1989). The town of Dingshu, and Yixing county, of which it is part, is one of the three main centres of ceramics production in China: Yixing and Canton are the two poles of earthenware and stoneware production, whilst Jingdezhen is the main centre of porcelain production. In 1958, an urban commune named the 'Capital of Pottery' (*Taodu*) was established in the township of Dingshu as part of Mao's ambitious (and soon to fail) Great Leap Forward, aimed at rapidly transforming China into an industrialized nation (see Jan 2004). The Great Leap Forward famously called for a rapid increase in steel production; as part of this ambitious plan, townships engaged in ceramics production nationwide became involved in a movement to 'use ceramics instead of steel' (*yi tao dai gang*), which involved establishing research centres to develop products such as machinery pieces that would otherwise have been made of steel (Dingshu 1992:20). After the failure of the Great Leap Forward, and the difficult years of the Cultural Revolution, Dingshu underwent a steady process

2 I use 'subjectivities' here in the Foucauldian sense of the 'subject' (see Julien and Rosselin 2009; Warnier 2009).

of industrialization, one which further intensified after the economic reforms of the late 1970s, and the subsequent economic boom in the 1980s and 1990s.

When I first visited Dingshu in 2003, I was struck by seeing pottery literally everywhere: streets were lined with all sorts of ceramic objects, large flowerpots, vases, tiles, all stacked one on top of the other, waiting to be loaded onto trucks or barges for countrywide distribution or export. Spatially, there was no distinction between residential areas and factories. Many factories were situated in the centre of town, and a stack of pots lying on the side of a street betrayed the existence of a factory hidden behind an apartment building. Historically the town owed much to the presence of the canals that criss-cross it and allow the movement of commodities to Shanghai and Hangzhou, from where they are also dispatched overseas. Until about the mid 2000s, the transportation of pots by barge was common; when fully loaded, these vessels would barely remain above water from the weight of the cargo. Ceramics production extends to the surrounding villages, both in factories and ubiquitous home workshops, typically made up of a team of husband and wife, and perhaps a son or daughter or other family member. In and around Dingshuzhen, more than 12,000 workshops produce *zisha* ceramics of all types and qualities, which leads Shi Juntang, head of the Yixing Ceramics Industry Association, to estimate that with an average three people working in each workshop, there are about 40,000 *zisha* producers in Yixing county.[3] If one were to include all related trades, including mining, clay preparation, transportation and so on, this number would be in the region of 80,000 to 100,000 people (Cheng 2010).

In 2003 and 2004, during the time of my main period of fieldwork, air pollution also attested to the title Capital of Pottery. Because so many of the kilns were still fired using coal, ash as well as clay dust was everywhere, and walking in the streets on a windy day could become particularly unpleasant. In the apartment building where I lived for the duration of my stay, a thin layer of black dust reappeared the day after the floors were cleaned. I was told that air pollution due to burning coal was at the root of a high incidence of lung cancer. I did not witness Dingshu at its worse, since many coal-fired kilns and factories were in the process of being destroyed. I was told that some years back, Dingshu had over 50 factories and over 100 chimneys spewing smoke from as many kilns. By the time I returned to Dingshu in 2006, government policies relating to air quality had forced factories to replace all coal-fired kilns with gas-fired ones, which meant an important improvement in health and environmental conditions. In the centre of town, high-standard apartment blocks and shops

3 Personal communication, 23 May 2012

had replaced old factories, and parks had been created along the main canal where in the past stood vast quantities of pots awaiting transport.

Such initiatives, no doubt manifestations of the newly acquired wealth of the town derived from its booming ceramics industry, are also linked to hopes for the development of the town for tourism: the region is famed for its tea, stalactite caves, nearby bamboo forests and, of course, ceramic products. China is rapidly developing, and I had the chance to witness what rapid growth meant in the case of Dingshu on another two return visits, in 2008 and 2012. By 2008, many of the artisans I knew had purchased their first car – in many cases luxury cars, including Land Rovers and BMWs. Tearing down of older factories and kilns does not, however, imply that production has slowed down in Dingshu. On the contrary, and true to its epithet as China's Capital of Pottery, ceramics production is booming to meet the demands of the rapidly growing Chinese middle class, and the production of *zisha* pottery in particular has undergone a further boom since approximately 2008.

Ethnographic fieldwork in the capital of pottery

My first encounter with *zisha* ceramics was in a small shop selling Chinese fine crafts and furniture, run by a Vietnamese of Fujianese origin in my home town of Geneva. I soon became intrigued by these small teapots made in an impressive variety of styles, mainly because they did not quite suit my idea of what Chinese ceramics should look like: although using motifs such as bamboo, chrysanthemum or prunus flowers that clearly identified them as Chinese, these pots were unglazed and simple in form, in contrast to the porcelain and other richly decorated ceramics most often associated with China. Beyond the aesthetic appeal of these objects, I was struck by the contrast, in my mind, between these traditional-looking pots, and the tumultuous history I knew producers had lived through. The question that came to my mind was: How can artisans who have lived through the Cultural Revolution produce such traditional-looking objects? This led me to think about the relationship between art or craft style and the social context in which it is developed, a theme that I thought would be interesting to develop as part of my doctoral studies.

In the case of Dingshu, the origin of this discrepancy between the traditional appearance of contemporary pots and the transformed social context in which they are produced might be self-evident. One can easily imagine that the first aim of the artisans of Dingshu is to produce pots that will sell, and sell for high prices, and understandably they will be catering to the tastes of their customers. The American artist Richard Notkin is known to have copied the form and feel of Yixing teapots to convey political and social messages – notably the dangers of atomic weapons – but artisans in Dingshu know that those who will buy their work, mostly businessmen and scholars in Taiwan, Honk Kong, Beijing or Shanghai, are looking for objects that are unmistakably 'Chinese' and 'traditional' in appearance. These craft objects can be presented as traditional precisely because they can be contrasted to something modern, and possibly evoke a feeling of nostalgia for a bygone era. But this fact in itself is interesting to address. Why attribute such a value to this aura of tradition in the modern world? There is also another series of questions that seems to be in need of addressing about the artisans producing such traditionalist objects: How can they claim legitimacy for a form of craft that the communist government attempted to transform into an industry, and after a period of history, the Cultural Revolution, that attempted to destroy such signs of traditionalism? And how do these claims to legitimacy relate to the transmission of craft-related knowledge and skills?

I arrived in Dingshu in September 2003, and was to spend twelve months in the town and in surrounding villages between 2003 and 2004, and pay shorter visits in 2006, 2008 and 2012. My first two or three months in Dingshu were a rather confusing experience. Things seemed to start very well: through an initial contact at the Chinese Academy of Social Sciences in Beijing, I met a native of Dingshu, a publisher living in Beijing, who proposed meeting me in her home town in order to introduce me to friends who might be of help. She arranged a meeting with Shi Juntang, head of the Yixing Ceramics Industry Association, based in Dingshu, who proposed introducing me to some of the most important masters of the town. This was an ideal opportunity, and occurred soon after arriving in the town. Over the course of two days, Mr Shi introduced me to six masters. In the interviews, I learnt about their work and style, as well as their experience of learning and working in the Zisha Factory, which was from 1954 to 1980 the only locus of production of *zisha* pottery. Mr Shi later invited me to a conference in town that brought together state officials and academics from several of the ceramics-producing centres in China. This gave me another dimension on craft production in China, one that took me a long time to fully appreciate: the state's strong involvement and interest in production, and in the crafts of the nation-state. It also pointed to significant overlaps between what in my mind were separate entities: crafts and industry.

After these initial contacts I started arranging some interviews of my own, either through contacts I made through Mr Shi, or simply by walking into workshops in town and in the surrounding villages. There I got a rather different picture of *zisha* craft, as well as the range of ways in which artisans worked and sold their work. In villages around the town, people produced cheap and roughly finished pots sold in tea shops throughout the country, and fired in impressive modern industrial-sized kilns within the village. In town, other artisans were making a living by making pots that would later be signed by some 'famous' artisan claiming to have made the object in question. More ambitious artisans, and those who could afford the extra time and money, were attempting to create a name for themselves by sending off their creations to local craft competitions. I got much information from these individuals about the social and economic contexts in which they worked, and I learnt about the workings of the *hukou* system, which had prevented rural dwellers from working in the cities since the late 1950s. Although this residential system had been significantly relaxed in recent years, it seemed to have imposed a clear contrast between rural and urban ceramics production.

My early interviews with artisans gravitated around the themes of the great tradition of *zisha* pottery, the prestigious history of the craft and the reasons for *zisha* pottery's uniqueness among the ceramic traditions of China. I found it more difficult to get artisans to speak about their daily lives, or to get a more precise grasp of the techniques of the craft. Paradoxically, I had access to many workshops but it was difficult to observe artisans at work since they would usually stop work when I entered to show me their produce; hoping for a purchase or feeling an obligation to serve tea to a visitor, at most they would briefly demonstrate one or another technique. Requests to film their work were usually met with resistance, or embarrassment.

Given these initial experiences, my first interview with master Ge Taozhong was particularly rewarding. Ge was an unusually quiet and very successful artisan in his mid forties, living and working in a house at the edge of the town, whom I had contacted through one of the masters I had met during my first weeks in Dingshu. He would answer most of my questions with his characteristic mildly ironic smile, and in fact soon interrupted our first interview that was being held in the sitting-room of his house to take me to his workshop in another room. There, he let me observe him as he went about demonstrating the various techniques of the craft in about half an hour, producing what I think of as a 'sketch' of a teapot: an object that, although not finished, contained all the elements that identified it as a teapot. This was the first time that I had been able to observe the techniques of *zisha*, save through glimpses in other workshops.

The next time I met Ge, I asked him whether I could film him at work. This he accepted, and over a period of two days I was able to document the entire process of the making of a *zisha* teapot, in this case a simple tangerine-sized one. This was another stage in my learning about the techniques of the craft, as I was able for the first time to observe the making of a handmade pot from start to finish. And for the first time, I began to grasp the many tricks and difficulties that the techniques of *zisha* involve. Ge was aware of this, and in the film he takes on the airs of a magician revealing his tricks. At an early stage of the process, after using a long, flat bamboo tool to detach a strip of clay from the table with a careful and expert hand, he looks up at me with a broad grin, then mimics the movement again to indicate that this was something worthy of notice. In another instance, Ge shapes the top and slightly concave part of the profile of a pot, looks up and indicates the curve of his own neck to point out how they are similar.

One of my intentions when arriving in Dingshu had been to learn some of the techniques of the craft in order to understand the experience of apprentices, and better understand these unique craft techniques. I had been intent on negotiating some kind of apprenticeship with a master, and meeting Ge, I hoped I could learn some of the craft's techniques from him. At first he seemed to be willing, though when I arrived at his house at the arranged start date, I realized he had other plans for me. He accompanied me to the buildings of the Zisha Factory and introduced me to his colleague, Zhao Jianghua, who was to become my teacher. Ge and Zhao shared the rent of a workshop, even though Ge rarely actually worked there. For about three months, I returned almost daily to Zhao's workshop to learn the craft. Three months is a modest period in comparison to the three years of a traditional apprenticeship (though many artisans claimed that six months were actually sufficient for a young person to learn to produce simple pots with the help of plaster moulds). The time I spent learning to make *zisha* pottery might have been too short to learn much about the technical difficulties of the craft, but I soon reached a stage where I was no longer learning new techniques, but practising what I had learnt, and I soon realized that this activity was taking time away from interviews and my exploration of other loci of production.

This period of learning brought much to my understanding of the craft, and the way it is taught in Dingshu. Beyond an understanding of techniques, I got a more direct glimpse of the relationship that is established between master and student in the context of an apprenticeship. Zhao was an ideal teacher, not only given his patience in explaining techniques and correcting my mistakes, but maybe more importantly for my research he was willing to share much about his own experience and impressions of the craft. This experience of learning the craft proved to be invaluable for me. As with many skills, those involved in the making of *zisha* might appear straightforward to the observer, but are in fact extremely complex to master, and this was an important reminder in order to understand the development of a career of an artisan in Dingshu.

My short 'apprenticeship' also proved useful in getting closer to the artisans of Dingshu, and an ideal complement to interviews and simple observation. Before starting to learn the craft I had felt that I was kept at a distance, but the news that a foreigner was trying to learn how to make *zisha* pottery spread quickly and became an ideal 'visiting card' as I went about meeting artisans. My master and others had not expected much from a foreign university student, probably thinking that I would give up after a couple of days, yet I believe they were genuinely surprised that I was not doing too badly. During my apprenticeship, I was able to get a feel of life in a workshop, the comings and goings of fellow artisans, the easy but at times tedious daily routine and a share of the local gossip. This was central in my thinking about the *zisha* 'community of practice', a theme I take up in Chapter 4.

Apprenticeship as a research method is in a way a particular form of participant observation, one which stresses the 'participation' side of it (Coy 1989a; Hsu 1999). Of important value is the knowledge that can be derived from apprenticeship. Most importantly, as in any craft or art, it is by trying oneself to create that one realizes the difficulties involved, and in consequence that one can better appreciate the skills of artisans. Since my research focused on a non-verbal kind of knowledge, namely the embodied knowledge needed to make a pot, apprenticeship was for me an important complement to the information I collected simply by asking questions. But there was another by-product of this involvement in the making of *zisha*, one that I had not anticipated: close to the end of my fieldwork, one of my main informants commented that I had become able to 'see' *zisha* pots, meaning that I was able to distinguish between a good and a bad pot. The word used here was *kandedong*, which might be glossed as 'understanding through seeing', and is a term that I heard repeatedly with reference to the craft. This capacity referred not only to an aesthetic sense, but also to one that was acquired through practice and involvement in the making of craft objects. That I was told this after several months of having observed artisans at work, talked with them about their craft and attempted a short apprenticeship myself was not fortuitous: this was a skill that I had started to develop only after extended exposure to the craft. This comment certainly pleased me when I heard it. I did indeed feel that I could see the pots in a certain way, and that I could recognize the signs of a high quality one. I had acquired this sense as artisans directed my attention to certain details while I learnt about the craft, either in discussions with artisans or during my apprenticeship. Evidence of this newly acquired capacity was my ability, at least to some extent, to predict an artisan's evaluation of a pot.

Back from fieldwork, however, I started reflecting on this phenomenon: How was it that I had come to judge pots in the same way as my informants? I felt that my fieldwork was an education of attention, a tuning of the senses, as described by the psychologist James Gibson (1979) and anthropologist Tim Ingold (2000), or Cristina Grasseni (2004), specifically with reference to vision: I had been taught not only to make pots but to appreciate their tactile and visual properties. But this education of attention had been associated with judgments of value: What makes a good or bad pot, and who is or is not properly following 'tradition'?

Susan Terrio (1996) has captured this phenomenon reflecting on her own fieldwork among artisan chocolate producers in Paris. Terrio had to prove her good taste – in particular, preference for dark over milk chocolate – to gain the confidence of her interlocutors. Citing Bourdieu's argument on the reproduction of class distinctions (Bourdieu 1984), Terrio discusses the use of an expert discourse borrowed from wine tasting among chocolate makers and connoisseurs:

> this oenological encounter powerfully illustrates how taste is produced and reproduced. [It shows how] consumers, even foreign anthropologists acutely aware of the processes at work, can be drawn into mastering, displaying, and ultimately, replicating taste protocols.
>
> (Terrio 1996:73)

I was taken in by artisans and their discourse about the impending disappearance of genuine 'traditional' craft. As a result, my research was biased towards that type of craft that claimed to be more traditional than other types, produced by artisans who might be identified as 'elite', those artisans who have gained recognition and were in a position to influence what was considered good and bad taste. I have since learnt to get some distance from some of the discourses I heard in the field, and have come to understand that it is not useful to dissociate those traditionalist elites from those who use non-traditional or modern techniques, or whose pots do not correspond to traditional patterns – all are taking part in the 'making' of a craft tradition and in the art world of *zisha* pottery.

Of course the situation I encountered is familiar to many ethnographers: as fieldwork progresses, one starts to get to grips with local politics, and appreciate the context of the claims of informants during initial interviews, to cast a critical outlook on discourse. This situation can take on a particular dimension for anthropologists involved in activities that mobilize the body, such as craft. Such fieldwork involves situations in which few words are exchanged over extended stretches of time, such as during an apprenticeship were the anthropologist is concentrated on what they are doing for long hours, with at times minimal interaction with others. As Wacquant (2006:69) reports of his experience as an apprentice boxer, fieldnotes have a tendency towards redundancy, mirroring the endless drill exercises he engaged in. What makes the power of informants' claims perhaps all the more compelling is that they refer to wordless and deceivingly 'natural' embodied processes (see Herzfeld

2004; Prentice 2008). In addition, there was a certain baggage I took with me to China when I set out to study craft: the ideas, dating back to the Arts and Crafts movement of the nineteenth century, that there is something inherently 'good' about working with one's hands, and that the work of the artisan is inherently satisfying. These ideas broke down over the course of my fieldwork, partly from talking with artisans who had become producers of *zisha* not out of choice but because they had been given their job in the context of a planned economy, or had more recently taken it up because of economic incentives. The lyrics of a traditional song that women potters would sing to make work more bearable claimed that they would marry the man who would take them away from the back-breaking work of potting.

Crafts and their politics

Arguably there are few nation-states besides China that had such explicit state policies throughout the second half of the twentieth century aimed at eradicating handicrafts as outmoded modes of production (Eyferth 2003; Schran 1964). The Chinese state under Mao sought to transform those crafts deemed valuable into modern industries, whilst artisans as a class were to be transformed into either peasants – if crafts had been a complement to farming activities (see Eyferth 2009) – or factory workers (Walder 1984).

Yet there are intriguing similarities with other histories of crafts in other part of the world, similarities that point to a need to understand the specificities of craft and artisans in order to better grasp their shifting roles and positions in the modern world. Certain themes crop up time and again in the anthropological literature on crafts from the four corners of the world: issues related to the transmission of knowledge, including apprentices being expected to 'steal' the knowledge of their masters (Herzfeld 1995, 2004; Marchand 2001; Prentice 2008); concerns about authenticity and the relationship between value and visual access to production (Dilley 2004; Gowlland 2009); and the recognition of masters as not simply proficient artisans but individuals with important social roles (Lancy 1980; Marchand 2001, 2009), These are all issues that are relevant in the case of China. This book attempts to chart the way issues of particular concern to artisans around the world take on particular meanings and expressions in the Chinese political, social and economic context of the second half of the twentieth and early twenty-first centuries, by investigating the processes that led to the construction of tradition, exclusivity and authenticity in craft objects in China following a period during which all expressions of the past were to be eradicated.

The production of *zisha* pottery, as with all aspects of Chinese social and cultural life, has been so deeply marked by the tumultuous events of the twentieth century that it would be difficult, if not methodologically flawed, to individuate the procedural and technical aspects of its production from the social environment in which it is produced. This means that methodologically, processes of knowledge acquisition or the sequence of making a pot should not be distinguished from, for instance, the particular cultural and political relationship between master and student, and aesthetic appreciation should not be held separate from the political process of the creation of mastery and legitimacy in speaking authoritatively about aesthetics.

I have found that the idea of 'craft' is useful not simply as a topic of research but also as a tool of enquiry. An 'anthropology of craft' necessarily runs into problems of definition: the term is an ethnocentric one, and usually used in the Euro-American tradition to indicate an activity or product that is neither art nor mass production. This is not a definition that can be exported wholesale to other cultural and social contexts. In characterising crafts, some scholars have variously focused on technique, as in David Pye's redefinition of craft as 'workmanship of risk' (Pye 1968); others in the Marxist tradition have been more concerned with craft as a mode of production (e.g. Cooper 1980), while those inspired by the Arts and Crafts movement have looked at the values associated with crafts (e.g. Betjemann 2008). What is so useful about 'craft', I suggest, is precisely that the term conjures at the same time values (the morality of making things with one's hands, of resisting capitalism and its tendency towards homogeneity), technical processes (how one actually makes objects, for instance a pot on the potter's wheel) and relations of production ('pre-capitalist', non-industrial and so on).

Thus, instead of dismissing the term 'craft' as problematic, I want to suggest it might serve the purpose of blurring registers, and enabling one to think across domains that tend to be treated separately in the social sciences. Whereas scholars have tried to identify and circumscribe what makes up the act of craft work or the relations of production that define it, I take the idea of craft as an ambiguous one, and useful for its ambiguity. Instead of identifying the essence of craft, I suggest we should explore the relational nature of what we consider 'craft'. Indeed, craft is always defined in terms of something that it is not, for instance something that can be contrasted with industry, mass production and factories, or art. Craft is also a confused and ambiguous concept in that it is used to refer alternatively to relations of production, to an activity or to a product. Craft objects are ideas as much as they are things, they are things at the same time as they embody values; as Webb Keane (2001) has beautifully illustrated, people might be suddenly reminded of the materiality of things that are supposed to have symbolic content, as when a cloth with symbolic value in a ceremony is torn: can the object still retain its symbolic content, or has it merely become a broken object? Rather than attempting to disentangle all the different facets of craft to identify its essence, I use the ambiguity of craft to show how these different facets of a single concept inform and influence each other. How is it possible, for instance, that similar kinds of work give rise to similar kinds of discourse about the value of that work? How is it that material objects that circulate between regimes of value create ideas about authenticity and heritage? How do technical processes give rise to possibilities for the division of labour and unequal work relations? Craft objects are not made in a political 'context', but are themselves political (Henare *et al.* 2007; Strathern 1990).

The difficulty in taking up such a perspective is avoiding the trap of what Jean-Pierre Warnier names the '*effet Magritte*' (Warnier 2009:25–6), which is the strong tendency in the social sciences when theorising about material culture of reducing things to their sign value. Warnier's solution is to think of the 'praxical value' of things: how the items of material culture that humans surround themselves with can easily become incorporated, part of the body schema and habitual movements of the body as if they were extensions of the body. He illustrates this idea with a conundrum that Mauss was faced with, when admitting in his essay on techniques of the body to not understanding how Berber men can run down sand dunes and not lose their slippers: Warnier (2001) suggests that these slippers have simply become an integral part of the person. Warnier (2009) goes on to apply this idea to an ethnography of a Cameroon Grassfields kingdom, and suggests that because material culture becomes part of habitual movements and thus 'naturalized', it can be usefully appropriated in relations of power. Thus, in Warnier's ethnography of a Grassfields king, an excess of implicit references to the body and things as containers justifies the rule of the king, who is himself a 'pot' that contains ancestral power that the king can disseminate to his subjects, for instance as he sprays palm wine from his mouth onto audiences during ceremonies.

This approach can be particularly useful in a study of craft to counter a tendency in phenomenologically inspired scholarship that assumes that engaging in the activities of one's informants, for instance through 'apprenticeship as a research method' (Coy 1989a), can give us some access to the subjective experience of our informants. Indeed, despite some very fine pieces devoted to the methodology of 'apprenticeship as method', a critical appraisal of the position of the anthropologist in such situations is rarely put forward (Prentice 2008 being a notable exception). A problematic consequence of the lack of critical appraisal of this approach, I would claim, is an assumption of a certain universalism of embodied experience, a claim Michael Jackson, a pioneer of the phenomenological approach in anthropology, has made: 'while words and concepts distinguish and divide, bodiliness unites and forms the grounds of an empathic, even a universal, understanding' (Jackson 1983:341).

In contrast to this approach, Warnier and other scholars of the French 'Matière à Penser' (MàP) group (Julien and Rosselin 2009; Warnier 2001, 2009) propose that we think of subjectivation, in the Foucauldian sense, in order to reflect on the experiential differences that divide and can lead to conflicts – as in the emotionally laden debates between hunters and environmentalists on hunting discussed by Warnier (2009:306–12). For Warnier and the MàP group, different habituation, determined by life histories of engaging with the material world, creates different subjectivities. What is missing in the phenomenological literature on skilled practice is the sheer political dimension of the work of the hands and engagement of the body with tools and materials that this literature tackles, the political dimension that constantly returns to mind in the context of fieldwork. In a study of craft, the MàP approach makes it possible to propose an analytical perspective that encompasses both the work of the artisan, consisting of a bodily involvement with tools and materials, and the social and power relations that define the artisan's place in a given community and society (see Naji 2009). In the following chapters, I consider some ways in which this occurs, including in contexts of learning craft techniques, interactions between peers, and between artisans and customers and the influence of the state on craft production.

Structure of the book

A number of themes relating to the production of ceramic objects in China are covered in this book. There are, however, two main lines along which the themes develop. A first line of enquiry follows the quality of craft objects. I discuss the phenomenon of China's Capital of Pottery, presenting the mesmerising volume of production of ceramics in Dingshu and surrounding villages. From a sketch of the production of low quality, mass-produced pots in the innumerable factories around Dingshu where I conducted fieldwork, I move on to the more specialized type of pottery that is *zisha*, and discuss the highly specialized and highly skilled handmade (as opposed to mould-produced) *zisha*. The latter will be my main concern, but I find it necessary to provide a context for the phenomenon of *zisha* production. The reason for moving from mass-produced commercial products to specialized craft is to evaluate the claim made by artisans that theirs is such a unique craft that few understand its methods of production outside the area, and few artisans are versed in the 'authentic' and 'traditional' methods of the craft. Out of the estimated 40,000 individuals making *zisha* pottery in the area, I was told only a dozen or so 'really know' the techniques involved in the production of entirely handmade *zisha* pots. How does one go from 40,000 to 12? In discussions of the 'purple sand' teapots of China in art and craft volumes published in English and Chinese, one is only given certain things to see: *zisha* is presented as an exclusive luxury kind of pottery, a tradition of making beautifully realized pots for which unique techniques are employed, and a craft with a prestigious history. What are the processes and strategies mobilized to make this craft so exclusive?

The second line of enquiry considers the path of becoming an expert *zisha* artisan, from apprentice to accomplished master. This is quite a broad scope for a single volume, and necessarily not all aspects will be treated to the same degree. Addressing all stages of the career of an artisan can, however, raise a number of insights that are not available in an enquiry that focuses on individual moments in a career. In particular, the two extremes of apprenticeship and mastery help me to situate artisans in relation to the objects they make. One could write about *zisha* whilst overlooking what happens in the first three years of the career of an artisan, namely apprenticeship. However, this is a stage when most of the principles of the craft are learnt, and the basic methods, repeated daily during the life of an artisan, are acquired. Conversely, masters are exemplar artisans, and a number of insights can be acquired as to the aspirations of artisans, and their assumptions about what constitutes good practice, good workmanship and a well-lived artisan's life. They are also individuals who have gained political influence within and beyond the craft community.

Chapter 2 introduces the main features of the craft, and contextualizes it in space and time. I discuss the phenomenon of the Capital of Pottery of China, presenting some of the artefacts that are produced in Dingshu, and consider the making of *zisha* from a historical perspective, looking at the transition from pottery as a family-based activity to a factory system in the mid 1950's, and the return of the family as an economic unit in China's more recent turn to capitalism.

In Chapter 3, I evaluate the claim made by artisans that *zisha* is such a unique craft that few outside the community of practitioners understand its techniques and methods. I consider the creation of value and exclusivity through interactions between artisans and their customers in the context of sales, and more generally as participants in an 'art world' in which discourses, writing and purchases contribute to negotiate the value of the craft in general, and of individual artisans and their work in particular. I show that artisans play a full role in this art world, not only in becoming connoisseurs of the craft themselves, but also in claiming agency and legitimacy in acting on the value of the objects they create. I suggest that the state plays a significant role in this art world in its efforts to promote the intangible cultural heritage of the country.

In Chapter 4 I look at the transmission of knowledge about the craft, from master to apprentice, and between peers. I consider whether one might speak of a 'community of practice' resulting from such interactions between experts and novices, and among peers. I look at the different apprenticeship systems – in the family, in the factory and in villages – and consider the interactions that exist, or are absent, between these spheres. I suggest that the creation of a cooperative factory has profoundly changed the features of apprenticeship and its social context, even after its ultimate privatization.

In Chapter 5, I continue with the theme of learning and apprenticeship, but propose a shift to the experience of learning and the intimacy of personal skills, whilst presenting the unique techniques that have been developed around a clay ill suited to the potter's wheel. I draw in part on my own experience as an apprentice to identify some of the central experiential and technical principles that inform the process of learning and making, whilst considering the consequences of the introduction of modern tools on the relationship between the body of the artisan, tools and materials.

Chapter 6 addresses a concept that figured prominently in the discourses of artisans, that of *fengge*, which might be translated as 'style', though the term will need to be qualified. I evaluate claims made by artisans that it is possible to recognize the maker of a pot, and in so doing look at the practice of signing one's work and at discourses on authenticity in which objects come to be associated with persons who might or might not be those who crafted them. I suggest that *fengge* is both the result of physical engagement with the clay and of other people's gaze. I go on to consider the significance of the concept for artisans, and in particular how it is used to chracterize their work as traditional as opposed to modern in the contemporary context of the revival of 'traditional' and recognisably 'Chinese' crafts.

Chapter 7 approaches the theme of mastery, and looks at the social and political conditions which elevate an artisan to the status of master. I examine the life of one individual, recognized as the greatest master of *zisha* craft in the twentieth century, whose influence some eight years after his death was still remarkably strong at the time of my fieldwork. The master is conceptualized in terms of a 'singularity' – an individual who, through personal charisma, is able to redefine the craft and its tradition. In broader terms, the chapter asks: What happens when creativity becomes entangled in relations of power within a craft community, as well as in the politics of the nation?

In the Conclusion, I bring together the different strands of inquiry to reflect on the relationship between creativity and negotiations of legitimacy and authenticity. This in addition enables me to reflect on the possible future of crafts in twenty-first-century China, and new relationships with the nation-state.

CHAPTER 2

Craft and industry in Dingshu

In this chapter I wish to contextualize the craft of *zisha* in space and time. More than simply giving a broader picture of *zisha* production, I want to show how it relates to the other forms of ceramics production that exist in the area, as well as investigate the relationship between cheaper, omnipresent mass-produced pottery and the finest *zisha* craft objects.

The origins and development of the craft of *zisha*, and in particular *zisha* teapots, are well documented, mainly due to the close links the craft entertained with the class of scholars since the Ming dynasty (Bartholomew 1990). Literati would comment on the work of particular artisans or on specific items, celebrate them in poems and reproduce them in sketches. The making of the first *zisha* teapot in the sixteenth century is in fact documented in the works of scholars of the time (Xu and Shan 2009:90–2). Supposedly, it was Gong Chun, the servant of a scholar visiting the region, who learnt the technique from one of the monks of a local temple, Gong Chun added a spout and a handle and transformed a common ceramic object into an item of value and prestige: a vessel that could be used for the preparation of tea, a drink highly praised by literati. A pot in the collection of the National Museum of Chinese History in Beijing is claimed to be that original Gong Chun teapot, though its authenticity is contested. Since this original *zisha* teapot, the craft has entertained strong links with the class of scholars of the area. Indeed, in contrast to other great ceramics traditions in China, such as the porcelains of Jingdezhen, the main patrons of the craft were scholars and not the Imperial court (see Lo 1988).

In addition to their use by scholars, *zisha* pots have also been common features in teahouses, and used by peasants working their fields. Proof of this is provided by artisan and local historian Pan Chiping.[1] Among his collection of antique *zisha* pots, three very different ones are signed using the same name seal of an eighteenth-century artisan: a large and roughly made and glazed pot destined for peasants (the base is blackened by the fire that served to reheat the tea during work in the fields); a medium-sized pot destined for the teahouses of the Jiangnan area, of which Dingshu is part; and a small, tangerine-sized pot destined for southern China, for the preparation of fine teas, such as the praised Oolong and Pu'er. *Zisha* pots were also an item of export, and were the first type of Chinese ceramics to be exported to Europe, initially used as ballast with the first shipments of tea, and later, in the late seventeenth century (Bartholomew 1990; Chu 1988), becoming a sought after type of vessel. The pots were also exported to Japan from the seventeenth century, and Thailand in the eighteenth century. Thus since the early days of its production, *zisha* participated in a global economy.

The craft of *zisha* is presented by its makers and connoisseurs as unique, starting from the very material it employs. *Zisha* clay contains a high ratio of iron (over 9 per cent), and as mentioned, it is said that the microporosity of the clay enables the tea to 'breathe' in the pot, fully releasing its aroma; further advantages are derived from its insulating properties, keeping the tea warm and preventing the pot from burning one's hands (Kerr and Wood 2004:275). Tea brewed in a Yixing pot supposedly tastes better than if made in any other vessel. This has been a received wisdom for tea connoisseurs in China for centuries, judging by the popularity of this type of vessel despite it not being the most elegant (especially in contrast to porcelain), and it has even been 'proven' scientifically in a study conducted by two *zisha* craft masters in collaboration with a university researcher (Tang *et al.* 1992). The study, also translated into English, analyses the chemical composition of tea brewed in three types of vessels – *zisha*, porcelain and glass – and concludes that tea brewed in *zisha* pots contains the highest levels of trace elements, vitamins and tannins.

Whilst the making of 'traditional' (or at least traditional-looking) *zisha* teapots is thriving, other types of ware have been the victims of modernization. Thanks to their efficiency, plaster moulds have almost entirely replaced the old-fashioned technique of beating clay, now only associated with the making of *zisha* pots, and the production of most other forms of ceramics – including *dagang* (large vats used for water and food storage), flowerpots, vases and cooking vessels – relies on the use of plaster moulds.

I introduce below the main features of *zisha* pottery, including methods of production and varieties of clay. Although concentrating on the modern era, I wish to consider the craft in a historical perspective to explain the

1 Personal communication, July 2004.

transition from pottery as a household-based activity (before the 1950s) to a cooperative and factory-based system (from the mid 1950s until the late 1990s), and finally to the current situation, which I think of as a hybrid of the two previous systems. But first, I wish to give an idea of the different types of ceramic objects, both *zisha* and those made of other types of clay, that are produced in the region.

Fine and coarse pottery

The pottery produced in Dingshu has long been classified as either *xitao* (fine pottery) or *cutao* (coarse or utilitarian pottery), referring to the type of object produced and its purposes. In the past, the production of fine and coarse pottery took place in different areas: the production of *zisha* fine pottery was mainly carried out in and around Shushan (Figure 3), whilst the production of *cutao* was concentrated in Dingshan – the townships of Shushan and Dingshan, together with the more rural Tangdu, merged to form the township of Dingshu in 1950. It was in Shushan that the first and second of the factories that produced *zisha* ceramics were constructed, in 1958 and 1980 respectively. Dingshan in contrast was known for the production of 'coarse' pottery, and notably of *dagang*, large multi-purpose and roughly made pots which still line the streets and canal banks of the town. Despite this distinction between fine and coarse, it was the production of *dagang* that was the most profitable industry in the nineteenth and early twentieth centuries. Other types of *cutao* items produced in the area included *shaguo*, earthenware cooking vessels with a lid and a single handle, flower-pots, and the *ong*[2] used for the storage of pickled foods and wine.

The kilns used until the mid 1950s were known as 'dragon kilns' (*longyao*), structures consisting of one single chamber, a long tunnel that climbed up a man-made hill, and could reach lengths of 70 metres. At night, the lights of the fire spewing from the side openings of the kiln recalled the image of a dragon, hence their name. In mid 1950s, the kilns were replaced by coal-fired ones, due to a shortage of the straw and wood that was used as fuel in dragon kilns. Only one dragon kiln remains active in a village near Dingshu. Estimated to be between 300 and 500 years old, the kiln is used to produce such items as flowerpots and mugs, but was never meant to fire *zisha*. Some *zisha* producers ask to fire their pots in the dragon kiln, which gives their wares an antique and unique appearance resulting from the resin in the pine wood used for firing, leaving residues on the surface of the clay. The kilns used for *zisha* and those used for *dagang* were different, and could not be used interchangeably given the different fuels used and the possibility that the glaze of the *dagang* could

2 Mandarin Pinyin: *weng*.

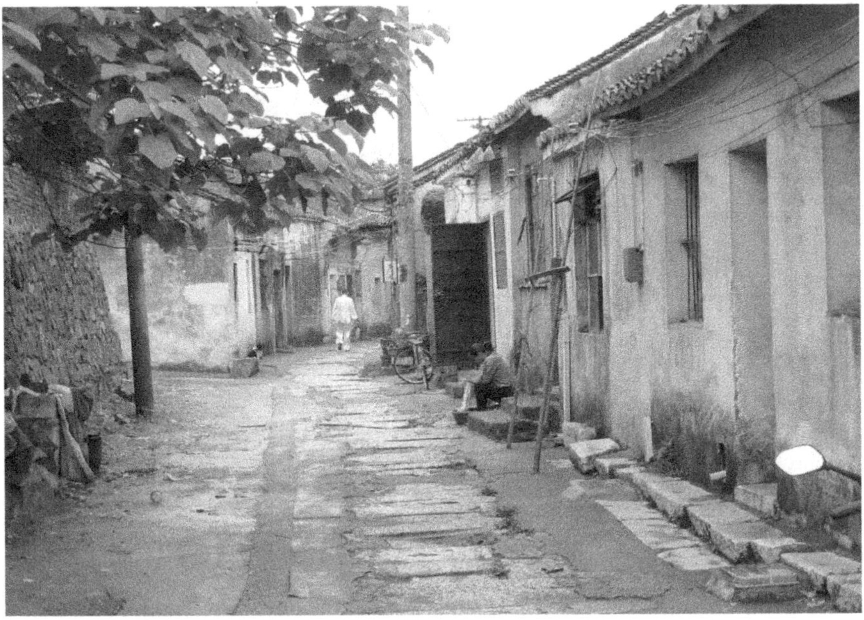

Figure 3 *Streets of old Shushan, one of the main loci of production of* zisha *until the 1950s.*

tarnish the unglazed *zisha* pots. The dragon kilns used specifically to fire *zisha* were called (perhaps confusingly) *da longyao* (Figure 4): large dragon kiln. 'Large' did not refer to the size of the products that were fired in those kilns, but to the height of the kiln tunnel in which saggars filled with teapots were stacked one upon the other up to a height of more than 2 metres (Xu 2009:112–16). Instead of readily available pine wood, which could tarnish the unglazed surfaces of *zisha* pots, the *da longyao* were fired using more valuable types of wood logged from surrounding hills, as well as straw. The original dragon kilns used to fired *zisha* were the first to be destroyed in the 1950s to make way for coal-fired kilns. The only example of an existing *zisha* dragon kiln is the one reconstructed on the grounds of master Xu Xiutang's house and company, built with the help of the kiln operators who once fired this type of kiln before they were destroyed. Between five and ten operators (*shifu*) were employed to conduct firing, the most important of whom was the *kanhuo* ('looks at the fire'), the person responsible for leading the operations by evaluating the temperature of the kiln according to the colour of the fire.

In dragon kilns, firing took two days. The process started with the slow heating of the kiln, with coal burnt in a furnace at its base. Kiln operators then went on to 'feed' the fire by pushing wood through one of a number of windows along the body of the kiln, and when sufficient wood had been inserted they went on to seal the opening with clay, and moved up to the next

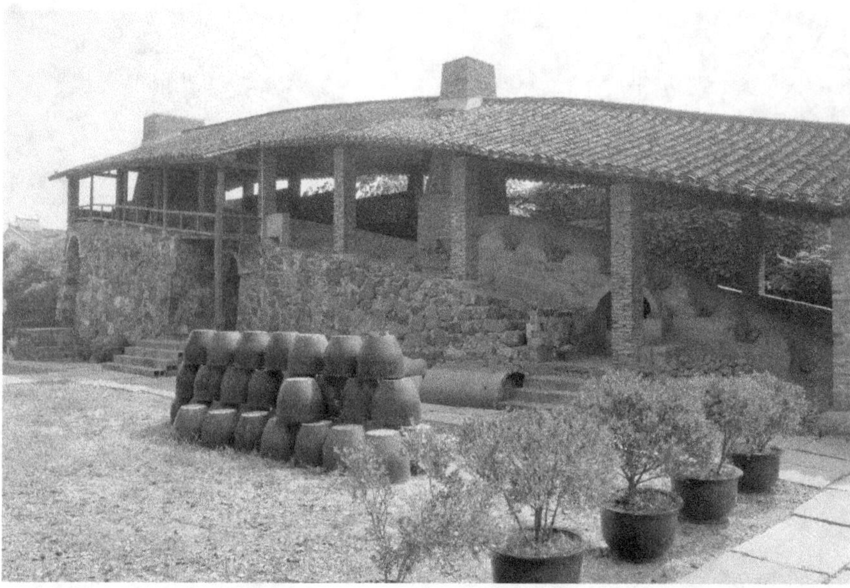

Figure 4 Da longyao *dragon kiln in the garden of Master Xu Xiutang. This is a reconstruction of the original kilns firing* zisha *that were destroyed in the 1950s to make way for modern kilns.*

window. Firing thus took place starting at the bottom of the structure, and slowly moved up until the final window was sealed, after which the kiln was left to continue firing its contents, and eventually cool down to a temperature low enough to be opened and unloaded without risk of thermal shock to the items it contained.

Dragon kilns were at first replaced by square coal-fired kilns in the mid 1950s – the first such kiln can be seen in the streets of old Shushan. This type of kiln was gradually replaced with more efficient ones. Firing of *zisha* today takes place in either gas-fired or electric kilns, though coal-fired kilns were still used into the early 2000s for other types of wares. Firing of *zisha* takes place at around 1200 degrees centigrade, which classifies the ceramics as stoneware. There are two types of kilns used for *zisha* today: the 'intermittent kiln', in which the pots are placed, raised to the desired temperature and cooled down before loading the next batch; and the impressive *suidao yao* ('tunnel kiln', also called *tuiyao*, or 'push-kiln') introduced in the 1970s, a kind of modern version of the dragon kiln consisting of a tunnel about 30 metres long, along which trays containing the items being fired are pushed forward by a machine. Firing happens in the middle of the tunnel, so that a single item will be slowly heated, fired, and have time to cool as it moves along the tunnel, eventually reappearing at the other end from which it entered. Standing at one end of the tunnel, it is possible to see the fire burning towards its middle and the profile of the pots being fired. This is a highly efficient system, which works day and night, and can fire thousands of objects every day.

It is the existence of such kilns, usually co-owned, until recently by whole villages, that has enabled entire villages to be involved in the production of *zisha* teapots (Figure 5). Artisans work at home, producing eight to twenty pots a day, depending on the intended quality and type. Then they take their work to the local tunnel kiln to be fired, paying a fee which amounts to around 1 yuan per pot, depending on its size. (This cost is approximately doubled or tripled in the closed type of kiln, in which temperatures are easier to control and better suited to the firing of higher quality items.) A disadvantage of tunnel kilns is the lower temperatures that can be achieved (which, depending on the product, might actually represent an advantage), and the fact that occasionally the trays containing the teapots tip and damage their contents. During a return visit to Dingshu in 2012, I was surprised to see that many villagers owned their own small electric kiln, which could fire only a few objects at a time. These had appeared alongside the rising trend in villages to produce more refined pieces, which might take a day or two of work per

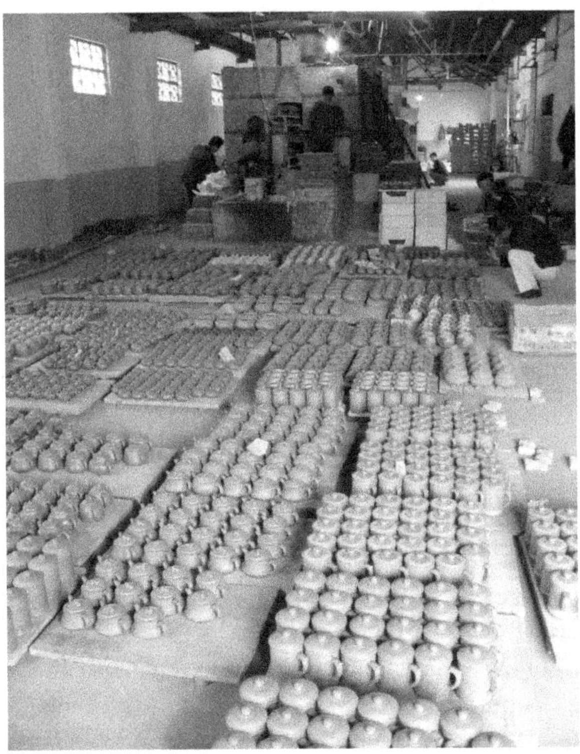

Figure 5 *Teapots waiting to be fired in the modern tunnel kiln of a village near Dingshu.*

piece. The risk of seeing such products damaged during firing justified extra investment in a kiln and the cost of electricity.

The characteristics of *zisha* clay

I mentioned in Chapter 1 that the clay used is said to be unique to the region. This is a matter of contention since some claim that the same quality of clay can be found in other provinces, but it is an idea firmly promoted by the head of the Yixing Ceramics Industry Association, Shi Juntang, who insists that only the clay of Dingshu has all the attributes that are desirable for the highest quality *zisha* pots (Cheng 2010). The most characteristic of the clays is *zisha*, though other clays, notably *benshanlu* (with yellow hues) and *zhuni* (red), are used, sometimes in combination with *zisha*, to create variations in colour. A common way for artisans to test the quality of the clay is to break a fired disc of the clay in two and touch the broken surface with their tongue: the artisan will know about the microporosity of the clay depending on how much the tongue 'sticks' to the clay.

Zisha, *benshanlü* and *zhuni* clays are not suited to being thrown on a wheel due to their high sand content, which does not make them elastic enough to be worked in that way. The techniques that have been developed to work the clay revolve around beating the leather-hard material into sheets and discs, which make up the different elements of a pot – a technique that will be more fully explored in Chapter 5. A similar technique, using the same beating method, was used until the 1980s, before the large scale introduction of plaster moulds and machinery, to make *dagang*. Methods of production at the time involved beating the clay into slabs, then wrapping these around a stone or wooden former, and pressing or patting them into shape against the former with a wooden mallet-like tool; the clay was then cut in two for the former to be removed, and closed up again with slip (wet clay used to seal joints).

The main difference with the techniques used in *zisha* is that the latter does not rely on wooden formers: the shaping of the body is done by holding one's hand inside the pot whilst beating the outside walls with a flat tool. Without the need to recover a former, which would get trapped inside the clay, the technique enables much greater freedom regarding the shapes given to a pot. According to *zisha* connoisseur Ip Wing Chi (1988), judging from shards found in the area, the technique involving patting clay around a wood or stone former dates back to the Song dynasty (960–1279). It seems that the technique used until recently in the production of *dagang* pots was the same used by Gong Chun to make the first *zisha* teapot (mentioned in Chapter 1). Indeed, a text of the time mentions that, in order to make the pot, Gong Chun first fetched a piece of wood that he had carved to the required dimensions in order to make his pot.

It was another artisan of the Ming dynasty (1368–1644), Shi Dabin (1573–1648), who first came up with the hand-patting technique (*da shentong*, or 'beating the body' of the pot). This was suggested in classic texts, and recently, Pan Chiping, a retired *zisha* artisan who has become a reference for historical research on the craft, claims in an article published by the journal of the Yixing Ceramics Industry Association to be able to confirm this (Pan 2003). His 'proof' is a teapot made by the Ming master, brought to Mr Pan by a collector for close examination. Pan concludes that the pot appears to be a hybrid of the two techniques of beating the clay around a wooden former and beating with the hand: the base of the pot seems to have been made by patting around a former, yet it seems that after the former was taken out, the hand-patting technique was used to close the upper walls of the pot. All other known pots made by Shi Dabin use only the *da shentong* techniques, and Pan suggests that this pot represents Dabin's search for an alternative technique. Despite a discourse that identifies the prestigious origins of the craft and its connection with literati and monks, historical research is thus revealing the more mundane roots of production techniques in the making of such common ceramics products as *dagang*.

Disappearing crafts

The making of 'traditional' *zisha* teapots is thriving. But it is other types of ware that have been the victims of modernization. Thanks to their efficiency, plaster moulds have almost entirely replaced the old fashioned technique of clay beating, and the *dagang*, flowerpots and cooking vessels produced in the factories of Dingshu are all made using such moulds or machines. People still remember the methods used by their parents or grandparents, and a certain nostalgia at times mingles with a certain pessimism as to the future of *zisha* itself. Not that the *zisha* pot will disappear – the craft is thriving as never before – but before long, some artisans claim, many of the age-old techniques of *zisha* will be forgotten. The plaster mould is also widely used in the production of *zisha* pots (Figure 6), and although artisans rely on many of the original techniques to make their pots rather than on the slip-mould method (used in mass-produced ceramics), and use the plaster mould only to perfect the final form of the pot, it is still seen as something of a 'degeneration' of the craft by those artisans who say that only an entirely hand-crafted pot can be made 'perfect' (*wanmei*), a theme I will return to in Chapter 5.

The relationship artisans have with the older techniques of *cutao* at times remains in the form of collecting. Xu Daming, a successful artisan, has built up a collection of Chinese antique ceramics, and he claims that some of his artefacts rival in quality similar ones collected by major museums. I was interested to observe how he handled some of the items in his collection. Xu

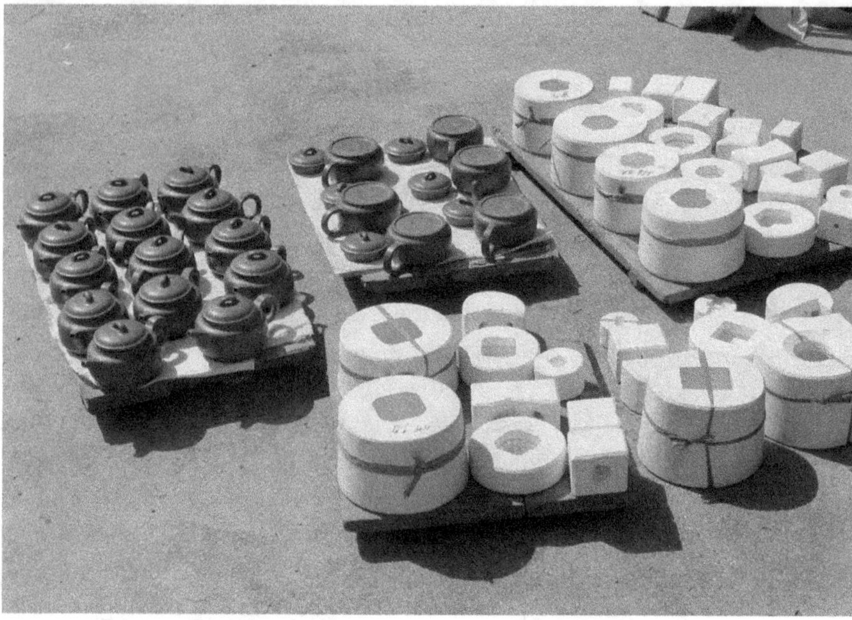

Figure 6 Teapots and the plaster moulds used to produce them lying in the sun to dry.

for instance explained how one item, a small lidded jar made of *zisha* clay, had been made: it had been shaped using the moulding method I have described for *dagang*, patted around a piece of wood or stone. Xu pointed out the lines on the body that revealed the way the tool had been used, and feeling the inside of the pot he could make out the joints where the walls met the part that formed the neck of the pot. Through such handling of antique objects, artisans become familiar with the way pots were made in the past, and might derive some inspiration from such objects to create new designs. Indeed, as Xu was showing me his collection, he made me to notice how he had reproduced some of the lines or decorative elements in his own work. In the specific case of this *zisha* jar, the rim is in the shape of a chrysanthemum, which is still a widespread motif often found in contemporary pieces with reference to this classic shape. The addition of such a motif however no longer has the function that it had in the past, namely of concealing the joint where the body and neck of the pot meet.

My teacher Zhao Jianghua explained to me that, in the case of one design of pot he often made, he resorted to a trick inspired by the mould-patting techniques he remembered his mother using when she made cooking vessels. In that design, one technical difficulty is the shaping of a long overhead handle (Figure 7). Zhao had devised a trick by which the clay was placed on a stone former he had had carved following the correct measurements, and patted into place with a bamboo tool in a way he claimed an artisan of old would have patted clay over a wooden former. With this 'innovation', Zhao had recuperated an element of older potting techniques, highlighting the relationship between fine and coarse pottery that has always been important, and still is according to Zhao, in *zisha* craft. This example brings up an idea strongly held among many artisans, that only innovations that reference tradition are acceptable improvements of the craft – a point I return to in Chapter 7. The status of objects produced by the parents and grandparents of contemporary artisans has changed from that of 'garbage' to antiques (Herzfeld 2004:196); Zhao displays behind glass in his workshop a roughly made candle holder that his mother used to make, and Xu displays old *dagang* pots in his large Zen-inspired garden.

Scholarship, consumption and export

Who were the purchasers of Yixing ware? The literature on *zisha* underlines the close links artisans had with the class of literati, who were also at times

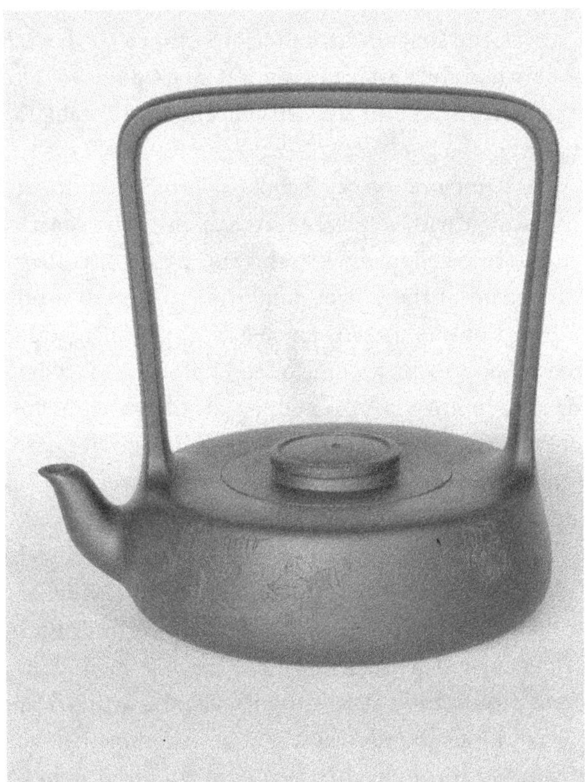

Figure 7 *Pot with overhead handle. This is a modern design by Gao Zhuang and originally produced by Gu Jingzhou. This pot was made by Zhao Jianghua, using the 'trick' described in the text to shape the handle.*

patrons, and had a tremendous influence on the aesthetics of Yixing ware (Lo 1988). Scholars were attracted to *zisha* pottery from the sixteenth century for its rustic appearance, which corresponded with their taste for rural retreats and ideals of simplicity: *zisha* pots came to be shaped in the form of natural objects – trees, vegetables, animals – and of antique bronzes and earthenware. A *zisha* teapot had to 'surprise, intrigue, provoke a smile, or excite erudition' (Brossollet 1998:7). It fast became an indispensable item in the paraphernalia of the literati.

The association with the literati was to take a new turn in the late nineteenth century, with the figure of Chen Hongshou, better known as Chen Mancheng, magistrate of Liyang district near Yixing county in the late nineteenth century and an enthusiast of *zisha* craft. Chen started a tradition of close collaboration between scholar and artisan: he is said to have designed eighteen kinds of teapot that he then commissioned from leading potters of the time. Museum curator Terese Tse Bartholomew talks of Chen bringing about a 'revolution in style' (Bartholomew 1990:46), which consisted in creating particularly simple designs with plenty of surface on which to inscribe poems and calligraphy. This was another major innovation to be brought to the craft: Chen was the first to attribute importance to the calligraphy and engraving that adorned the surface of a pot. Chen's pots would often include three signatures or seals: one of the artisan, one of the calligrapher and the third of the engraver.

Today, it is not uncommon for scholars, professors of art or artists to collaborate on a design with a talented artisan, as is the practice of handing over an unfired pot to be engraved by scholars, artists or fellow artisans. This act can raise the value of the object many times, even of a pot of relatively poor quality. I heard of one master, for instance, who helped his friends and students in times of need by inscribing a few characters of calligraphy on their pots. Collectors are interested in obtaining an object with the mark of the master, regardless of the quality of the pot itself. In another case, the recently apprenticed son of a master would have his pots inscribed by his father, a guaranteed way of finding purchasers of his work.

With the Republican era (1912–49), *zisha* pottery saw its traditional patrons, the literati, replaced by industrialists from Shanghai. Shops were established in Shanghai by powerful households from Dingshu, recruiting artisans from Dingshu and artists who would engrave the unfired pots. In such a context, the pot would have at least two seals, the artisan's and that of the shop, and the decorations themselves might be accompanied with the seals of both the engraver and the shop. Artisans continued the style of nineteenth-century production, and some of the most talented of the 1920s and 30s were asked to produce forgeries of classic pots to suit the high demand for antiques. Artisans were given at times months to work on a single item, in order to produce a most credible forgery. According to *zisha* artisan and local historian Pan Chiping, they derived their inspiration from reproductions in classic texts, actual pots found in private collections and at times their own imagination.[3] It is claimed that these forgeries were so convincing that the authenticity of antiques in contemporary museum collections is still difficult to establish today; one of my informants, an expert in the history of *zisha*, even claims that most items on display in the most prestigious museum of Hong Kong, the K.S. Lo Flagstaff House Museum of Teaware, are fakes, although of course if this is indeed the case, these forgeries would still be valuable as having been made by some of the masters of the early twentieth century. This has interesting repercussions on the idea of 'traditional' *zisha* pottery: quite literally, this is an 'invented tradition' (Hobsbawm 1983), where genuine 'classics' are indistinguishable from forged ones, and might all serve as reference points to present-day artisans wanting to reproduce a 'traditional' pot.

Although scholars and collectors where not the only users of *zisha* pots, one knows far less about other consumers, simply because they do not figure in the texts of scholars. This is the case, for instance, of pots made for peasants of the northern areas of the country dating from the mid eighteenth century. These were an altogether different type of pot, much larger, of a capacity of 1 litre or more, and rougher in appearance. Some *zisha* artisans might have such a type of pot in their collections, and from these objects one gets an idea of the uses these pots were put to: they are often blackened at the base, which indicates, according to Pan Chiping, that they were also used as kettles. The peasant would bring the pot full of water to the field, and later in the day build a fire to heat the contents.

In addition to domestic demand, an important percentage of *zisha* production was destined for export. There is a long history of export of *zisha* pottery to Japan (from the sixteenth century), and to South-East Asia (in particular Thailand) and Europe. As mentioned, *zisha* was the first type of Chinese ceramics to make the journey to Europe, used as ballast in the first shipments of tea in the late seventeenth century; early *zisha* pots destined for export to Europe are notable for their pronounced decoration, to suit the tastes of European consumers, and were sometimes embellished with silver and gold mounts by European metal smiths (Bartholomew 1990). Export to Europe was, however, not to last long, only about fifty years, as Europeans soon discovered another and to their eyes more aesthetically pleasing type of Chinese ceramics: porcelain. Yet Yixing ware (*zisha*) had an important influence on European ceramics design: what is known as Boccaro earthenware, for instance, was originally a copy of *zisha* ware made by Cleffins in Holland in the eighteenth

3 Personal communication, July 2004.

century. In England, the Elers brothers in Staffordshire produced their version of *zisha* teapots, as did Böttger in Germany, around the same time (Brossollet 1998). Soon, Europeans no longer needed to import its *zisha* pots, as they could produce them at home.

From kiln households to cooperatives

In the production of *zisha* ceramics in the Dingshu area, access to and ownership of kilns established hierarchies between families. Kiln households (*yaohu*) were mostly same-surname groups that owned sections of dragon kilns and either purchased unfired ware, or greenware (*pei*),[4] from rural producers or rented out kiln space to smaller households (*xiao yaohu*, 'small kiln households'); the latter were usually attributed the sections of the kiln with the least efficient firing, at the top and bottom of the kiln tunnel. Producers of greenware were at the bottom of the hierarchy and were paid by the piece by kiln households. Other households were involved in the extraction of clay and excavation of new mines, in the procurement of fuel and in the transport of products on barges (Xu 2009:15). An 'official history' of *zisha* pottery, compiled by managers from the Zisha Factory and included in the Jiangsu province gazetteer of 1989 (Anon. 2012),[5] suggests that the system of producing *zisha* described here, with property-owning kiln households purchasing unfired ware from rural producers, consolidated itself during the late Qing dynasty (which ended in 1911). At the time, there were no guilds for producers of greenware, but an association of kiln households was established in 1915 to regulate the affairs of the craft and set prices.

In Dingshan and Shushan, the largest producers of pottery before 1949 were not involved with *zisha* but with the production of the large *dagang* vats. Two lineages, the Ge and Bao, were the most powerful in the Dingshu area and controlled the production of *dagang* in Dingshan (Anagnost 1985:33): they owned thirteen of the twenty-odd kilns associated with this trade, and the ancestral temples of these lineages were supported by the sale of digging rights in the clay pits they owned. These families extended their business through

4 The word *pei* is usually rendered in Pinyin as *pi* in dictionaries, but I have stayed true to the local pronunciation of the word.

5 This text, circulating on the internet, purports to be a history compiled by the managers of Zisha Factory with a view to inclusion in the Jiangsu provincial gazetteer of 1989. Only fragments seem to be included in the gazetteer of 1994, which I have consulted, and I was unable to get hold of the 1989 version. An informant did suggest that he recognized at least parts of that text as genuine. I use this version of the history of *zisha* as it is richer than published gazetteer versions. I use references to this text with caution, and only in cases where it confirms details that I have heard from informants.

joint ventures in Japan, and were also the ones to receive most government support during the Republican era to build their businesses. In her study of the ceramics of Dingshu, Ann Anagnost (ibid.) notes, however, that given risks of firing, family fortunes could be rapidly reversed, and within these powerful lineages, kiln rights of poorer households could pass to more successful households, forcing the poorer ones to rent out kiln space from others. This meant that the division between kiln households and skilled pottery operators was not always clearly marked.

In contrast, there were no such large households in the production of *zisha* pottery. No household owned an entire kiln; rather, these were part-owned by three to four smaller kiln households. Anagnost (ibid.:27–8) suggests that producers of fine (*xi*) pottery, in other words *zisha*, were the most 'proletarianized'. Because no one owned an entire kiln, households would take turns to fire their wares in the different sections of the kiln so that all could take advantage in turn of the central part of the kiln where the most efficient firing took place. Those households that had part ownership of the kilns would employ up to about twenty workers, and have a shop front in the streets of Shushan. Given their relatively small size, the kiln households involved in making *zisha* were not stigmatized as exploiters after the founding of the People's Republic in 1949; the class label *zhong yaohu* (middle-sized kiln household) was applied to such households, a label roughly equivalent in Maoist classification to upper-middle peasant classes, but not to landowners, the latter being equivalent to *da yaohu* (large kiln household) (ibid.:36).

During the Sino–Japanese War (1937–1945), production of *zisha* pottery came to a standstill. The invading Japanese destroyed seven dragon kilns, and close to 100 workshops, stealing or destroying tools and machinery. Between 1937 and 1940, none of the remaining kilns were in operation. Production resumed when the local warlord forced kiln households to produce for export to Japan, but these activities ended in 1943. Victory over the Japanese in 1945 did not mark an immediate return to production; on the contrary, it presented kiln households with a problem, since it meant the complete loss of the Japanese market on which they had relied heavily in the years preceding the war (Anon. 2012).

Cooperative and factory: 1952 to 1979

In the initial years of the newly established People's Republic, the government claimed ownership of the kilns of the large kiln households, who had been impoverished as a consequence of the war years, and requested members of these households to collaborate in the context of the joint (state and private) ownership of production. Members of *da yaohu*, notably the Ge and Bao lineages, were stigmatized after 1949, regarded as exploiters; as mentioned, the label *da yaohu* was more or less equivalent to being characterized as landlords, which they also often were (Anagnost 1985:37). Yet my informants insisted that they played a decisive role in the early years of the People's Republic through the joint private and public ventures, which were needed by the new government in the reconstruction process and in the initial transition to collective ownership given these households' knowledge of kiln operation and management. The state also took control of the extraction and distribution of raw materials, and the first state-owned factory in Dingshu, established in 1955, took over all clay mines in the area and unified extraction operations, training and distribution among the factories of Dingshu. One department took over the fifteen clay pits on Huanglongshan where *zisha* clay is found, regrouping several clay-extracting households that previously owned them. In 1958, the factory introduced innovations such as electric drills and mills, and introduced deep-shaft mining (Xu 2009:16–17).

In 1952, responding to the drive to collectivization, workers in Tangdu, which had been part of Dingshu since 1950, created the Tangdu Cooperative, producing *ong* pots used for the preservation of pickled vegetables and wine. In 1954, the Shushan Cooperative was set up in Shushan, the traditional centre of production of *zisha*. The cooperative was established with loans from the then successful Tangdu Cooperative, and comprised three workshops, two of which made *cutao* products, and the third, the Zisha Workshop (Zisha Gongchang), resumed the production of *zisha* after years of interruption. An artisan who joined the Zisha Workshop in 1954, together with his parents and brothers, explained to me the relief this meant for him and his family, who all produced greenware (and as such had been a particularly poor household): they could then eat their fill of the meals provided by the cooperative, and they were guaranteed a constant wage rather than being dependent on the whims of the kiln households to whom they sold their unfired pots. In 1956, production output of the various forms of ceramics surpassed that of 1936, which had been the high watermark of production before the war, and meant that the devastation of the Japanese invasion and ensuing years of civil war until 1949 had been overcome (Anagnost 1985:29). In the Zisha Workshop, each year for the first three years about twenty apprentices between the ages of 14 and 16 were recruited. During an apprenticeship which was meant to last three years (but would be interrupted in 1958), they learnt the theory and practice of *zisha* pottery, and were taught history and theory of art in the evenings.

The government aim in encouraging the creation of cooperatives was to integrate the handicrafts (*shougongye*) of the country within the socialist project. Handicrafts, to the Chinese communists, were seen as 'a traditional and technically outmoded method of production which combines labour with a few simple tools and implements without the aid of mechanical power' (Schran

1964:152; see also Toops 1993). The 'modernization' of handicrafts involved their gradual absorption into modern industry, a process that, in the same way as agriculture, was to be at first self-financed rather than relying on state support, hence the formation of cooperatives and cooperative factories. The latter were to be upgraded to state ownership when the conditions were deemed 'ripe', that is, after the mechanization of production (Eyferth 2003, 2009). Artisans were a politically suspect class in the eyes of the new communist government, since they owned their means of production, and were therefore prone to the 'spontaneous, large-scale reproduction of capitalism and the bourgeoisie' (Liu, quoted in Eyferth 2003:55). Their status was thus ambiguous: they rendered a useful service to the nation as 'labouring people' (*laodong renmin*), yet they were not yet 'workers' (working class, *gongren jieji*) (Eyferth 2003:55–6).

In an undated catalogue of products for export from the early 1970s that an informant had kept in his drawers, an introduction to the craft of making *zisha* pottery, translated into English and referring to a Maoist slogan, paints a positive picture of the transformation of the craft by the communist government and suggests that it would have died out under the former government, the Guomindang (Kuomintang):

> The production of Ihing Tzu-Sha [Yixing *zisha*] earthen ware has a long history. As far back as in the years of the Ming Dynasty, it had developed fairly well.
>
> During the reactionary rule of the Kuomintang, imperialism from without and feudalism and bureaucrat-capitalism from within subjected the industry to untold devastation that it was brought to a dying state.
>
> In the course of ten odd years after liberation, under the leadership of the Chinese Communist Party and the guidance of its policy of literature and art of 'letting a hundred flowers vie in beauty and letting the new evolve out of the old', the industry has gained its speedy rehabilitation and flourished resulting in the expeditious growth both in the excellence of quality and the volume of output. The increase of the product year after year in export has made its appearance in the markets of over 30 countries and regions and enjoyed the acclaim of a broad section of foreign consumers.[6]

In 1958, in response to the Great Leap Forward, the Zisha Workshop of the Shushan Cooperative was transformed into a factory, the Zisha Gongchang, or

6 Undated catalogue published by the China national Arts and Crafts Import and Export Corp., Kwangtung Branch. No page numbers.

Zisha Factory;[7] that year, a middle school was established as part of the factory. The creation of schools within factories was part of the communist project of mass education within the newly founded communes, a project aimed at producing a population of literate and technically trained workers and peasants over a short period of time, consistent with the Great Leap Forward's ambition of rapidly transforming China into a military and industrial power (Jan 2004). Over the following two years, students were recruited by the hundreds, so that in 1961, the factory had over 2,000 workers and over 300 students. The apprentices who had joined the Zisha Workshop since 1954 were given the duty of acting as mentors to the new recruits. The latter came not only from Dingshu, but more distant villages. One artisan mentioned that his brother-in-law came from a village 30 kilometres away from Dingshu, looking for work. He said that most of those who had heard the news that the Zisha Factory was recruiting joined without even knowing what it produced, and many in fact thought it was involved in clothing manufacture because of the reference to the colour purple with the word *zisha*.

At this time, the Zisha Factory started to diversify production, and introduced slip-casting techniques, used in parallel with the more traditional techniques of *zisha*. Workers were recruited from increasingly different backgrounds; whereas in the first years of the factory students were locals and came from families connected with the ceramics industry, from 1958 onwards, recruiting was extended to young persons from the rural areas. One artisan, Mr Fang, joined the factory in 1958 at the age of 14. He came from a large kiln family, and his parents considered the production of unfired ware as a lowly activity, yet had to accept that their son take up such a job due to the difficult economic conditions at the time. After the failure of the Great Leap Forward, and following new government directives aimed at limiting the rural exodus (the *hukou* system of residence and work permits), the factory once more changed its recruiting policy in 1961. All those who had an agricultural background were sent back to work in the fields, and Zisha Factory was left with between 600 to 800 workers until the mid 1990s.

The status of the Zisha Factory throughout its existence was that of *da jiti*, or a large collective. It never acquired the sought after status of *guoying*, or state-owned factory (unlike other ceramics factories in Dingshu). For this reason, it was considered of lower status to work there, and workers did not enjoy all the social security benefits reserved for workers in *guoying* enterprises. The difference between *da jiti* and *guoying* is, however, one of degree rather than kind, given that *da jiti* factories, in the same way as *guoying*

7 Confusingly, the words for 'workshop' and 'factory' are pronounced in the same way, but written differently, in Chinese. See glossary.

factories, are thoroughly embedded in the state administrative structure (Yang 1989:41) and rely on the state sector, for instance as a source of capital, materials and equipment, and sales revenue (Jefferson and Rawski 1994:61).

Some technical innovations were brought to the craft with the creation of the Zisha Factory, including the modernizing of the kilns, a shift in fuel from straw and wood to coal, mainly due to shortages, and innovations in clay extraction with the establishment of deep-shaft mining (operations carried out by another factory based in Dingshu). The techniques of the craft remained, however, to a large extent unchanged. Although the introduction of plaster moulds in the first years of the factory represented a major technical innovation, the mould was adapted to the craft rather than the other way round in the case of the production of *zisha*. Indeed, initial experiments revealed that *zisha* clay is not suited to the method of slip-casting usually used with a plaster mould.[8] Instead, something of a hybrid technique was created whereby clay was beaten into sheets and discs, as with previous methods, and then pressed inside the mould to give it its shape. The method helped to give the pots a more even shape in a shorter time, although for those artisans already well-versed in the original techniques, moulds did not necessarily speed up the process. In fact, the greatest advantage of moulds was that they guaranteed a much shorter period of learning: at the time of the Great Leap Forward, driven to account for the efficiency of the factory, a Zisha Factory slogan proclaimed that new recruits could be transformed into fully skilled workers within three days – a claim that, in line with communist modernization guidelines, contributed to portraying the making of *zisha* as no longer an 'outmoded' craft but increasingly a modern industry. The three days of apprenticeship were, however, only put into practice to train workers using slip-casting techniques, a parallel production line in the factory.

Another paradox in the 'modernization' of the craft lies in the fact that the factory did not attempt to introduce a full-scale division of labour. Although, as in the past, clay preparation and kiln operations were carried out by specialized workers, artisans continued to produce pots from beginning to end, and would even make their own bamboo and wooden tools. Perhaps surprisingly, apprentices who spent up to one year learning to prepare the 'building blocks' of *zisha*, namely the sheets and discs of clay, did not pass these on to more skilled workers to be transformed into pots. In other words, no 'vertical' division of labour (Goody 1982a) was introduced, whereby the

8 Slip-casting consists of pouring liquid clay (slip) into plaster moulds, which absorb the excess water forming a 'crust' on the walls of the mould. The excess slip is poured out after ten to thirty minutes, and the clay shape that is made is left to dry before being removed from the mould.

less experienced, or apprentices, carried out simpler tasks to improve the efficiency of the masters' work.

Beating sheets of clay was not the only simple task that could have been accomplished by apprentices: preparing slip (fairly liquid clay used to join pieces together), or beating unused clay for it to be reused, were all rigorously carried out by individual artisans. I was given the explanation that a master cannot use the clay prepared by a student, since the quality of every stage of the process is reflected in the finished object. The logic seems to be that a chain is only as strong as its weakest link: a master's pot using clay prepared by an apprentice appears to be the work of an apprentice. I would also add another technical reason for not delegating simple tasks to apprentices, namely that an artisan must be aware of the stage of dryness of the clay at any moment, since certain actions can only be performed at specific stages, for instance the clay must be worked on soon after it has been beaten into sheets.

The factory seems therefore to have relied very much on the principles that regulated *zisha* as a craft, without attempting to introduce significant technical changes. Jacob Eyferth (2003) describes a similar situation with paper-making in Sichuan. Traditional methods did not benefit from economies of scale, and the small production units entered the factory unchanged. However, Eyferth points out, efficiency of production was not the aim of the government in introducing such factories; the issue was rather one of control: 'collective workshops, like collective farms, did not improve productivity but facilitated supervision' (ibid.:65). But the slow pace towards innovation and labour-intensive production can also be explained pragmatically as an attempt to minimize costs, and the choice of certain techniques of production depends on relative availabilities of factors of production (Riskin 1971). In the production of *zisha*, the factor of relative availability was people with an already acquired set of skills. According to an informant, in the early years of the Zisha Factory, plaster was more expensive than flour and used minimally, as it was not cost effective.

Over the period of activity of the factory, from 1954 to 2002, the focus of production shifted repeatedly. In the early years of the cooperative and factory, the focus was on the production of wares similar to those produced during the Republican era: items of daily use, often decorated, and principally aimed at export. Significantly, one of the masters recruited by the cooperative in the first year of the factory was an engraving and calligraphy specialist, and he and his students engraved or otherwise decorated the pots of the other artisans working in the factory. In successive years, the factory had to readjust its production line. During the early years of the Cultural Revolution, it shifted most of its production to items destined for the masses, including large, roughly made teapots to be used by peasants, responding to government directives to orient industry towards agricultural needs (cf. ibid.).

Consequently, the production of more traditional *zisha* teapots stopped, as they were considered one of the 'Four Olds' that the Cultural Revolution was supposed to destroy. At that time, throughout the country, shops selling *zisha* teapots were forced to sell their stock at a loss. Items were now engraved with quotes from Mao or other phrases that promoted the ideals of the Cultural Revolution, for instance the phrase *dahai hangxing kao duoshou*, 'relying on the helmsman when sailing the ocean', the title of a song that promoted the thoughts of Mao Zedong (Xu 2009:80). During the Cultural Revolution, the Zisha Factory also trained peasants in the rural areas around Dingshu to produce part-finished (unfired) products using moulds designed and made by the factory's workers, and fired in its kilns. Production at the factory also varied to include industrial wares such as water-filtering appliances. A short-lived experiment in techniques also led to the use of screen-printing appliqué techniques for printing effigies of Mao on cups, though the results were not ideal and abandoned soon after (ibid.:80). It is also worth noting that from the very start of the factory there was a section employing sculptors, devoted to the making of busts, including of Lenin and Marx, using *zisha* clay.

Of course, the Cultural Revolution was also marked by political tensions throughout the country. The campaign to 'destroy the Four Olds' did not only affect the production of wares deemed traditional: in 1966 in Dingshu, more than 100 clay and wood figures of gods were seized and brought to the worker's stadium to be destroyed. The same year, rebel factions – which were encouraged by Mao in order to act as a counter-force to local officials and criticize cadres – were established in schools and factories, and fighting between factions led to violent and sometimes deadly clashes in the factories and streets of Dingshu until 1969, when they were disbanded, also putting on hold production for extended periods (Chen 1992:26–7).

The making of more recognisably 'traditional' *zisha* resumed in 1970, when items were once again produced for export. Leafing through a copy of the previously mentioned catalogue of the factory's products for export, one artisan, Yi, was able to point out those pieces that he recognized had been made by particular masters, and those pieces that clearly had been made by hand, without the use of a plaster mould. He explained that some designs, as I have mentioned, were faster to make by hand than by mould by those artisans who had learnt the traditional methods of making *zisha*. The catalogue also shows a few innovations that had been made in the design of new wares, including, intriguingly, a coffee set aimed at foreign markets. Other pieces are more readily recognizable as *zisha* ware, reproducing traditional motifs such as bamboo trunks, prunus and so on.

Dingshuzhen and Yixing county since 1979: the turn to private enterprise

The favourable economic climate created by the 1979 economic reforms meant opportunities for increased sales and the creation of new markets for *zisha* teapots, enabled by the entrepreneurship of overseas businessmen. In particular, a Hong Kong businessman and Yixing ware enthusiast, K.S. Lo (in Mandarin, Lou Guixiang), did much for the revival of the market in fine *zisha* teapots in Hong Kong and Taiwan. According to one artisan and former manager of the Zisha Factory, who remembered K.S. Lo's visit in 1979, Lo asked masters to reproduce masterpieces from the Ming and Qing dynasty, using copies of pieces from the Shanghai Museum; these items were destined for Honk Kong and South-East Asian markets. A few years later, in 1984, Lo set up a museum of Yixing ware in Hong Kong. The museum regrouped both antique and contemporary pieces from Lo's collection, which remains to this day one of the richest collections of *zisha* pots. Through his activities, not only did Lo renew interest in the finely crafted pots, but he also contributed to the reintroduction of the practice of signing pieces, which had been discontinued during the years of the Cultural Revolution (items were then stamped with the characters *Zhongguo Yixing*, 'China, Yixing'). Overtaking Hong Kong, Taiwan soon became the largest market for *zisha* ware. Some artisans recall how, in the early 1980s, traders from Hong Kong and Taiwan would buy pieces from famous artisans even before they had been finished or fired. According to informants in Dingshu, this enthusiasm was driven by the desire of wealthy Taiwanese to create their own tea ceremony (*chadao*).

Although it was illegal to do so, due to tense relations between China and Taiwan, a number of Taiwanese traders travelled to Dingshu in the years following the Open Door Policy of 1978 to purchase stocks of pots for resale to Taiwanese consumers who were eager to purchase such prestige items. In this way, Taiwanese traders bypassed those from Hong Kong who were making profits from the demand for *zisha* pots in Taiwan. The Taiwanese traders risked years of imprisonment by doing so; arrangements were made between Taiwanese and mainland-Chinese ships to meet at sea in the middle of the Taiwan Straits in order to ensure passage of the Taiwanese to the mainland.

Walking in the recently constructed ceramics market in Dingshu during a return visit in 2012, I came across a sign on a shop door claiming that they were selling pots bought by the Taiwanese in the 1980s and that 'returned' to the mainland. The owner of the shop, a Taiwanese himself, explained that the tables had turned, and it was now mainland Chinese who had become wealthy and desiring of prestige objects such as fine *zisha* ceramics. The added value of 'Taiwanese-bought' pots is not to be underestimated: it is a claim to authenticity, in more ways than one. Pots produced in the 1980s were strictly produced in what became known as the Number One and Number Two factories, which guaranteed to some extent the quality of the wares and of the clay, which is often said to have been better (rightly or wrongly) to present-day clay. Of course, it is not always easy to determine whether the pots were indeed purchased by Taiwanese in the 1980s. The trader explained that this could be shown by judging the quality of the clay, and demonstrated by pouring hot tea in the pot, pointing out that the colour of the body of the (hot) pot is slightly darker than the (still cool) lid. This, according to the trader, is caused by the particular porosity of the pot, which has absorbed the water and has become darker as a result.

One of my informants later offered a slightly different explanation to this change in colour: it does not indicate the relative quality of the clay used but rather the temperature of the kiln at the time of firing. A slightly lower temperature will achieve this porous result, and the tunnel kilns used until the 1980s by the factory, because of their nature, fired at lower temperatures. In contrast, modern gas and electric kilns are used at slightly higher temperatures. Some masters use lower temperatures to achieve the same effect as the earlier firings; the resulting pots are less shiny than the pots fired at higher temperatures, and as a result might make less of an impression on customers at the time of purchase. However, connoisseurs will appreciate that the pots are seasoned more easily (by pouring hot tea over them), and will acquire the sought after sheen of continual use more rapidly than pots fired at a higher temperature. Thus, the change in colour of the pot when hot water is added, as demonstrated by the Taiwanese trader, might indeed be an indication, but not a guarantee, that the pot was made in the 1980s. The Taiwanese trader claimed to have another way of ensuring that his pots were genuine: when he bought them from Taiwanese collectors, he did so on condition that he could get a refund if he got evidence that the pot was fake. He then travelled to Dingshu and asked the sons, daughters or students of the presumed master to establish the authenticity of the pot. (I return to the topic of authenticity, and such measures to distinguish fake from genuine items, in Chapter 6.)

With the economic reforms of 1979–80, and with the opportunities provided by the boom in demand for *zisha* ware, four other factories were created in the early 1980s under the system of privately owned township and rural factories. In 1980, the first of these factories was opened, opposite the original Zisha Factory. It was named, and is still known as *Zisha Er Chang*, 'Zisha Number Two Factory', while the original factory became *Zisha Yi Chang*, or 'Zisha Number One Factory'. It was initiated by a township cadre, Shi Juntang (who would later head the Yixing Ceramics Industry Association), who called on artisans and managers at Number One Factory to teach new recruits, and was soon in competition with the original factory. Over the following three years, rural *zisha* factories numbered three to five appeared in surrounding villages.

The 1980s were marked not only by structural but also substantial changes in *zisha* production: artisans started regaining a degree of freedom. The first *zisha* factory created a research department (*yanjiusuo*), which consisted of several individual workshops where masters, with one or two apprentices, developed new designs, created high quality pots and taught advanced students. With the demand for *zisha* still high, artisans recalled how they would return home in the evening to continue work illegally. In the factory, artisans were paid by the piece, but they could get three to five times more money per item if they sold their work furtively to traders, without the intermediary of the factory. Private production and direct sales proved for many to be more lucrative than the factory system, so that soon those artisans who could afford it set up a workshop in their own home, and sold directly to traders.

In 1996, the workers at Number One Factory had no other choice than to do likewise or find other work. Starting in the early 1990s and within the framework of its economic reform plans, China intensified the restructuring of its state-owned enterprises, disbanding those that were not turning a profit and removing barriers that formerly prevented competition between state-owned enterprises, cooperative enterprises and foreign imports. Although not under state ownership, large cooperative enterprises were thoroughly enmeshed in the state system and dependent on other state enterprises, and the economic reforms also had consequences for these (Jefferson and Rawski 1994). In this particular context, the managers of the Zisha Factory stopped providing a guaranteed income and benefits to its workers. In effect, the latter were laid off though they were allowed to continue using their workshop space to produce ceramics (though many left to work from home), use the factory's kilns (at a price per piece) and rely on the factory for sales, for which it charged a commission.

With this new state of affairs, the majority of workers were unable to compete and forced to find other sources of income. Renewed pressure followed soon after, with the crash of the Asian economy in 1997, which affected Taiwan, the major market for *zisha* pottery. It was around 2003 (the year of my first visit to Dingshu) according to my informants that the demand for *zisha* pottery started to pick up again, mainly fuelled by the growing mainland Chinese middle class. The original Zisha Factory remained a cooperative until 2002, when it was privatized and became the Yixing Fangyuan Zisha Arts and Crafts Company (Yixing Fangyuan Zisha Gongyi Youxian Gongsi). The passage from public to private property did not imply major changes beyond those introduced in the mid 1990s. The company rents out about 100 individual workshops as well as kilns, clay processing equipment and shop spaces selling *zisha* pots.

Craft and the Chinese state

The state was obviously very present in the affairs of *zisha* production during the years of socialism, and in particular from 1949 to 1979. This state presence is not entirely new, although one gets the impression, reading about the history of *zisha*, that it was relatively limited in the past. Scholars were the main patrons of the craft (Lo 1988), and state presence has, for instance, been far from as important as in Jingdezhen, China's 'Capital of Porcelain', where imperial kilns had been managed by government officials since the Yuan dynasty (Gillette 2010; Huang 2008). In Dingshu, *zisha* was a relatively marginal form of ceramics compared to other more important products such as *dagang*. There were some limited connections with the imperial court: *zisha* ceramics were sent to the Imperial palace as tribute during the reign of Kangxi (1661–1722), and during the reign of Qianlong (1735–1796) *zisha* teapots were acquired for use at the Imperial summer residence (Anon. 2012). New techniques were developed to distinguish these imperial wares, including enamelled decorations and intricate surface carvings.

When the state became interested in supporting ceramics production in Dingshu during the Republican era, most efforts seem to have been diverted to the production of *dagang*, and it is this investment that transformed the Ge and Bao lineages into particularly powerful kiln households, enabling them to extend their business to the Japanese market. In the case of *zisha*, limited efforts at building the industry resulted in the creation of a township government-sponsored factory in Shushan, which produced *zisha* pottery alongside other types of ceramics, and trained new recruits; it might in fact be regarded as the first government attempt at creating a school for teaching the making of *zisha* (ibid.). Another effort followed that same year with the creation of a *zisha* department within the Yixing county vocational school devoted to 'kiln professions' (*yaoye*).

During the Maoist era, beyond the policies that affected the recruitment of workers, and life and work in the factories, another policy, the *hukou* system had strong repercussions on Chinese handicrafts by regulating who did and did not have access to work in factories. The *hukou* was a system of residential and work permits essentially aimed at slowing down the rural exodus. At first intended to be a short-term measure, it eventually lead to the crystallization of geographic divisions through the restriction of mobility (Cheng and Selden 1994). Every citizen was given a certificate indicating their occupation, social class and geographical origin (peasant or urban worker; inhabitant of a medium, large or very large city). Crucially, the *hukou* system regulated access to work in the cities – peasants were not allowed to join urban factories and were to remain peasants.

The implementation of the *hukou* system provoked a forced urban exodus, whereby those considered peasants were sent back to rural areas, with the effect of marking more sharply than before the division between town and country. The sociologist Norman Stockman (2000:46) argues that the *hukou* system was the last element that served to congeal the separation between urban and rural society, a process that had begun in the late nineteenth century with the development of the economy and of technology, and intensified when the communists took power. According to Stockman, before that time China did not have any structural or cultural division between town and country, yet the opposition between city and countryside was central to the social analysis of orthodox Marxism, and it was thus necessary that it was created. In Dingshu, since 1961, work in the urban factories was restricted to those classified as residing in urban areas, and the still apparent division between the township and surrounding villages, in terms of production and wealth, has its roots in the *hukou* system.

State influence and involvement in the affairs of the craft has taken several other forms since the founding of the People's Republic. I explore some of these in later chapters, including in relation to the politics of heritage (Chapters 3 and 7), titles and competitions (Chapter 6) and vocational education (Chapter 4).

Rural production and migration

During the 1960s and 1970s, as mentioned, peasants were not entirely excluded from the production of ceramics: the Zisha Factory had a monopoly not so much over the production of *zisha* pots but over their firing and sale. Parts of the production process were transferred to the surrounding villages: the factory sent workers to teach pottery to villagers to give them a second source of income. The villagers would work at home and take their unfired pots to be fired in the factory's kilns. Factory workers would select the good pots, smash the faulty ones and pay the villagers a few mao (tenths of a yuan) for the good pots.

It was only in the early 1980s, with the possibility of establishing rural factories, that members of the rural community became once again actively involved in the making of *zisha* ware, though access to Number One Factory was still restricted. Later, from the 1990s onwards and with the creation of privately owned kilns, it became possible for villagers to work at home and have their work fired locally. *Zisha* is particularly suited to home production: it is made with few and readily available tools, and only a small workspace is needed. Some of the villages around Dingshu resemble factories in themselves: between one and three kilns might be constructed and held in co-ownership by two or three bosses (*laoban*), who together will have invested approximately 150,000 yuan in an average kiln. Access to kilns is open to the community; artisans rent a slab for between 8 and 10 yuan on which can be fired a dozen or so pots. Other smaller, closed kilns are also owned privately, and rented out for the firing of more prized items.

Today, the line of separation between the town of Dingshu and its surrounding villages is no longer enforced, even though the *hukou* system and some of its restrictions remain. Villagers are allowed to work in town, and some artisans living and working in the villages enjoy to some extent the fame and popularity of their town counterparts. Yet I often found that in discourse, the division between town and country appears to be maintained. In particular, the urban/rural dichotomy takes the form of an alleged superiority of urban artisans, with particular reference to those employed by the original Zisha Factory. For an artisan, being able to mention in their c.v. that they worked in the original factory is greatly advantageous, and part of the *zisha* economy of fame. Therefore, it should not come as a surprise that the most famous artisans working in Dingshu started their career there. Today's greatest masters are those who were among the first recruits of the Zisha Factory.

Moreover, the separation between country and town is often recalled with reference to presumed differences in style (*fengge*) between villagers and townspeople. The pots made in villages, I was told by townspeople, are rougher (*cu*) in appearance than their urban counterparts. On this basis, some artisans claimed to be able to recognize whether a pot had been made by someone from the town or from the villages. The idea that artisans in the rural areas should make rougher-looking pots also reinforces a stereotype of them having little understanding of the craft. This was used to discredit their work, but also at times, to imagine a more genuine, less intellectualized relationship to the craft.

The division that separated the town from the villages has been supplanted by another type of division in recent years, between locals and migrant workers. Factories producing other types of mass-produced wares – such as roof tiles, flowerpots and jars – which use low-skill techniques such as slip-casting or shaping bowls and plates by machine (jiggering and jolleying), tend to employ mainly migrant workers. Southern Jiangsu is one of the richest areas of China, and as a result its inhabitants can easily find better sources of income than working in local factories, which has created a vacuum filled by migrants from other parts of China. The division between *zisha* and these other wares is mirrored by that between 'locals' (*bendi*) and migrant workers, coming from poorer provinces such as Anhui, Sichuan, Guizhou and even the northern part of the Jiangsu province, much poorer than the south. The division between locals and migrants is however rapidly changing: during my last visit to Dingshu in 2012, increasing numbers of migrants seemed to be involved in the production of *zisha*, or were involved in activities related to the craft, including clay preparation.

Relations of production: makers and designers

I refer to producers of *zisha* as 'artisans', yet that label can often be misleading. Indeed, many of those who present themselves as artisans (*taoyijia*) might more correctly be identified as entrepreneurs or merchants. Thanks to plaster moulds, it is possible for an individual to compose a teapot design without having to know *zisha* techniques: the pot is 'designed' by selecting and combining different plaster moulds for the different elements that make it up – body, spout, handle and so on. The moulds might be made from a clay model, though most of the time they are simply machine shaped, and different motifs are sold as standard. Moulds are produced by other artisans according to the instructions of the designer, which might include an object to be reproduced or a photograph or drawing. The mould so obtained is then passed on, often together with the clay, by the entrepreneur (referred to as a 'boss', *laoban*) to a worker (*gongshou*) who will use the mould to create a clay object. The entrepreneur will also give their own personal seal to be stamped on the base of the pot by the worker. This means that the worker is denied the possibility of associating their name with their creations.

This system, referred to as *jiagong*, can be described as a classic 'putting out' mode of production, whereby a merchant supplies part of the materials and tools to the worker and pays upon collection of the final product. In Dingshu, however, there is a twist to this classic model: in many cases, the merchant presents themselves as an artisan, the creator of the objects on sale, attested to by the seal stamped on the base of the pot. The 'merchant-artisans' in these cases are locals. In villages, it might be local entrepreneurs who hire fellow villagers to produce teapots bearing their name. Many established artisans also rely on the system of *jiagong*. One rather successful young artisan, in his late twenties, explained to me that he found making pots rather tedious, whilst he enjoyed designing new works. Although he was perfectly capable of doing the work himself, he had decided to ask others to make pots from his designs. For *jiagong* workers, this practice is also often a way to start a career, relying on the success of other artisans before working independently and signing their own work.

The system of *jiagong* is particularly widespread in the villages. Entrepreneurs are individuals who have been able to build their fame through the promotion of their work in competitions, and in many cases by claiming a tie of kinship with a famous master. Some villages, however, have acquired enough of a reputation for artisans to be able to set up their own shop, and use their own seal to stamp their wares. One of these villages is Shuangqiao, which is reputed for its production of small pots of relatively good quality. In Shuangqiao, eight out of ten households are engaged in the production of *zisha* ware. Here, the main street is lined with houses separated into living quarters, on the first floor, and a workshop, on the ground floor; the workshops also act as sale points.

During visits at the time of my first period of fieldwork in 2003/4, I found that many of the artisans were married couples working together in one workshop. One artisan explained that it was mostly individuals between the ages of 30 and 50 who were making *zisha* pots, since older individuals had not learnt the skills and continued to work the land if still able to do so, whereas younger people were not interested in taking up the craft, preferring to find more secure employment in factories. This claim was disproved when I returned to Shuangqiao in 2012; taking advantage of the resurgence in demand, artisans of all ages were producing *zisha*. In Shuangqiao, villagers started working in this way in the early 1980s. Before that time, they made flowerpots that were sent to the Zisha Factory to be fired; at the time of my original fieldwork in 2003/4 they could rely on their village's three privately owned kilns, and when I visited in 2012, many had purchased their own electric kiln, deemed safer than the tunnel kilns of the village.

The gendered division of labour

Much was done by the communist government after 1949 to improve women's condition, such as participation in schools and factories, and many contemporary *zisha* masters are women who were able to join the cooperative and factory soon after it's establishment. Women's participation in the production of *zisha* does not date from this period, however: in the past, women as much as men where producers of *zisha* pottery. Women were primarily involved in the production of greenware, whilst the activities linked to clay extraction and kiln operation was the domain of men. Women rarely received recognition of their work before 1949, though some craftswomen had gained fame in their lifetime. The first craftswoman to be mentioned in the contemporary scholarly literature is Yang Fengyeng, born in 1796, who, together with her two brothers gained a reputation for producing pots based on the designs of Chen Mancheng (Baidu Baike n.d.b). Another woman who gained a degree of recognition before the communist era was Jiang Rong, who was based in Shanghai in the 1930s and was involved in the reproduction of *zisha* antiques. Jiang Rong was the only woman master to join the *zisha* cooperative when it was founded, and she is recognized as one of the greatest masters of *zisha* pottery of the twentieth century, second only to Gu Jingzhou. Today, another craftswoman, Wang Yinxian, has become the figurehead of *zisha* pottery, the only artisan named as an 'inheritor' of the craft within the framework of the national list of intangible heritage (see Chapter 7).

It is common to see husband and wife working side by side in the workshops of Dingshu, as well as in the nearby villages. Working in tandem has a number of advantages, including the possibility of sharing tools, materials and workspace. Husband and wife work individually, each with their own design of pot. Among artisans with significant reputations, I have noted that men tend to be the ones who take more credit for their work and attain a higher level of recognition, as the development of the husbands' careers often takes precedence over that of the wife, who is more involved with domestic tasks. In contrast, during interviews, Wang Yinxian stressed the sacrifices that her husband – an engineer in another factory in Dingshu – had to make for her to develop her career.

CHAPTER 3

Creating value in the *zisha* 'art world'

The original Zisha Factory, which started as a cooperative in 1954, is now famous for having employed, since its creation, most of the greatest *zisha* ceramics masters of the twentieth century. The buildings of the now privatized factory are a tourist attraction, and individual or groups of (mostly Chinese) tourists occasionally visit the factory compound. Most visitors will stop in one of the ceramics shops there to purchase some souvenirs on their way to another attraction in the area; unlike so many other shops around town, these offer a certain guarantee regarding the quality of their products. A few tourists might, however, venture into the buildings, going from workshop to workshop to discuss purchases directly with the producers. Mr Yi has his workshop in one of the buildings of the former Zisha Factory. This ten-floor building, constructed in the 1990s, houses about seventy private workshops, in which usually one to three artisans work. Yi usually leaves his door open, as many in the building do, and visitors usually feel free to walk in uninvited, often without announcing their presence with a knock on the door, nor acknowledging the presence of the artisan. Yi would usually not interrupt his work, nor a chat with a colleague or with me, to interact with a tourist, an attitude which I often found disconcerting. Yi has, on one wall of the workshop, a showcase displaying some of his works, a short c.v. printed on cardboard and a small collection of various items, including an old teapot that would have been used by peasants working in the fields in a not too distant past. On more than one occasion, I observed how tourists would look at Yi's works on display with a feigned knowledgeable air, and ask about prices. Yi, without lifting his eyes from his work, would say matter of factly, 'eight to ten thousand yuan', to which the tourist would give a quick laugh (which I interpreted as half embarrassment and half scorn), put on an air of knowing exactly why these pots were so highly priced (tourists would claim that they could buy excellent quality pots for a fraction of the price in another workshop in the same building) and promptly leave.

Yi's attitude in the presence of a potential customer reveals much about relations between artisans and their customers in Dingshu. That Yi does not encourage sale is understandable, as he knows perfectly well that no casual tourist is ready to pay the prices he asks. But I am interested here in the relationship that is established between an artisan and a potential customer. What is happening in such kinds of interactions? Here the valuation of objects steps over into the mutual valuation of persons: as soon as a price is revealed, a relation is established between the maker of expensive objects and the tourist, who is not only taken aback by the price, but crucially incapable of 'seeing' the worth of the pots.

A casual visitor to Dingshu is confronted with a vast array of prices, and no doubt is often confused by it, as I often was when trying to buy pieces for my own collection. Two similar pots can be sold for very different prices, depending on who made them or, more precisely, on the maker's claims – to have been the student or nephew of some master, for instance, or have received numerous prizes in competitions. I mentioned in the previous chapter that *zisha* pots come in many qualities, yet disparities in quality only begin to explain the enormous disparities in monetary value between different objects. In the villages, a trader might buy quantities of pots for a few yuan each. At the other end of the spectrum, masters sell their work for several dozen if not hundreds of thousand yuan: the record for a contemporary master's pot has been set at 300,000 yuan (about $45,000) at auction. Value might be in part determined by the material employed – websites on *zisha* connoisseurship abound with discussions about the dwindling stocks of 'real' *zisha* clay – but this only starts to explain the disparity of prices. What then makes a *zisha* teapot valuable?

To answer this question, in this chapter I address the craft in terms of an 'art world' that sets in place 'rules' for the valuation of individual items and of the craft as a whole. More broadly than simple monetary value, I want to show how cultural and aesthetic value is created in objects. I argue that artisans are not passive but fully agentive in these valuation processes. Roy Dilley (2004) has pointed out that anthropologists have tended to concentrate on value creation in the context of exchange, following Appadurai's notion of the politics of value (Appadurai 1986). More recently, consumption seems to have become central to anthropological analyses of value. Dilley tackles the question of the relationship between consumption, demand and the 'politics of value' on the one hand, and production – relations and modes of production –

on the other. He suggests expanding Appadurai's concept of regimes of value to include not only the 'pathways and diversions of objects through social contexts, mediated by social structures, but also the relations and processes of production that give rise to the objects so exchanged and consumed' (Dilley 2004:797).

This will be the topic of this chapter. Any discussion of the value of *zisha* ware would be incomplete without mention of the relevance of *zisha* as a collectable item. This discussion will entail situating *zisha* items in the wider context of Chinese culture, in reference to which, I maintain, *zisha* pots act as miniatures (Lévi-Strauss 1966:22–5). However, the attribution of value to objects is not only linked to their meanings, but also to subjective feelings – such as nostalgia – and ambiguous and problematic evaluations such as authenticity. I explore these themes throughout the book; in this chapter I focus on interactions between artisans and their customers in the context of sales, and more generally on artisans and connoisseurs as participants in an 'art world' (Becker 1982; Moeran 1987). In such contexts, artisans and connoisseurs, through discourses, writings and purchases, negotiate the value of the craft in general and of individual artisans and their work in particular. Further than claiming legitimacy in the evaluation of their own craft, artisans have developed a discourse in which they claim that connoisseurs can rarely fully understand the craft. I will suggest that artisans claim full agency and legitimacy in acting on the value of the objects they create. In order to set the framework of the discussion, I start by identifying the consumers of *zisha* craft.

Consumers of *zisha*

Artisans would often talk to me about their customers, trying to determine their demands, concerns and tastes. If, as Dilley (2004) has pointed out, the relationship between production and consumption has not been made clear in anthropology, one way of addressing the question is to look at the discourses of producers about what is perceived to happen at the moment of purchase and consumption. Approaching this issue in such a way acknowledges the agency of artisans in acting on sales rather than passively responding to demand for their products. Artisans pride themselves on having a very precise idea of the tastes of their customers. From the vantage point of Dingshu, artisans classify Chinese customers according to tastes in tea and teaware, drawing a mental map of places which will be the targets of their production. China is seen as composed of three distinct areas, corresponding to different preferences in tea drinking and, as a consequence, different sizes of pots. This imagined map was often recounted to me during my fieldwork; indeed, it was part of a repertoire of the main particularities of the craft, and likely to be told to any newcomer to Dingshu. Artisans say that a teapot has to adapt to the person who uses it, because in every part of China people drink tea in a different way, corresponding to different habits and customs. In northern China, I was told, people like their teapots large; medium-sized teapots are made for the Jiangnan area (south of the Yangzi river), in which Dingshu is situated; and southerners (including the Taiwanese) like to use small pots, the size of an orange or sometimes even smaller. Explanations differed in a few details, but it basically revolved around the claim that in northern China, the diet includes a lot of onions, which are supposed to make people thirsty. Also, northerners usually like to drink their tea in large quantities in one sitting because of cold weather during the winter, so they prepare the tea once and in a large teapot. In the past, the main users of *zisha* pots were peasants, who would take their teapot filled with tea with them to the fields, and needed large pots to have enough to drink throughout the day. In the Jiangnan area, it is warmer and drinking too much hot tea in one go would make people sweat, so the tea is sipped throughout the day instead. In the south, teapots are used to prepare *gongfu* tea, or teas that require careful preparation to bring out their full flavour – by definition, expensive teas. They are also prepared very strong – so the teapots need to be small, to economize on use of the precious tea, and are used to prepare just enough for appreciation rather than to quench one's thirst.

Artisan and local historian Pan Chiping started collecting antique *zisha* pots when he retired; his collection privileges historically significant rather than beautiful pieces. He is particularly proud of three pieces representing the three *zisha* 'regional' prototypes, which he likes to display side by side to show visitors. It is intriguing that these three items, dating from the eighteenth century, carry the same seal: they were made by the same artisan. Artisans at that time did not specialize in a given size of pot, but adapted the objects to their customers, resorting to this three-fold division of China and Chinese tastes in tea. Another artisan, Ge Taozhong, had spent a year in a technical school in the nearby town of Wuxi to learn industrial design. There, he had been introduced to the European aesthetics of the Bauhaus and its ideas about beauty deriving from functionality. In a conversation with him, I suggested that *zisha* pottery might be compared to industrial design for deriving beauty from the function of the object. However, Ge pointed out what he believed to be significant differences: industrial design makes objects that are universal in aspiration. For instance, it designs tables that are to be used by all those who want to use a table. So the designer has the idea of an object, and sets out to design what they believe is the best version of it. Conversely, the *zisha* artisan cannot make universal items, they must know the tastes of their consumers, whether they will want a small or large pot, and even what kind of tea will be brewed in it. Ge told me of one of his customers in Taiwan who asked him to make five identical teapots using five different clays, so he could decide which clay is best suited to which particular tea. Ge says that as a result of all these interactions, he has learnt to appreciate tea, and has a good idea of the type of pot he is to make when connoisseurs approach him.

The threefold division of China according to tea preferences seems to be less important today than in the past. To begin with, customers' profiles have changed. The large pots made in the past for peasants in northern China are no longer produced – they have been replaced by cheaper and more readily available vessels, such as the Thermos flask, glass or plastic mug. The threefold division has however been replaced by other ways of categorizing customers. I was told for instance that Chinese mainlanders prefer highly decorated wares, whereas Taiwanese and Hong Kong people like simpler wares. Tastes in tea still influence the size of pot, notably in the case of drinkers of expensive Pu'er, produced in Yunnan, and Tie Guan Yin, produced and consumed in Fujian province and Taiwan, for whom smaller *zisha* pots are the only vessels suited to their preparation.

Artisans are also aware that different areas have different 'typical' customers. In particular, among mainlanders, those who can afford the high prices asked by masters are often politicians and/or government or high-ranking military officials. One artisan once told me he sold his work for higher prices on the mainland to politicians who did not, he said, understand the craft but were willing to pay high prices for such prestige items. His Taiwanese customers, however, tended to be scholars or connoisseurs of fine craft and tea; they were less wealthy than mainlander customers, and because they understood the worth of the works, the artisan quoted cheaper prices to them. (Left unsaid was the likely fact that, as a connoisseur, a Taiwanese customer was more likely to talk about his purchase, maybe even in a connoisseur's publication, which would bring fame to the artisan.) Many artisans today realize that their pots will not be used to prepare tea at all, but be collected as works of art. This has contributed to a diversification of forms, since a teapot need not even be required to make tea in. These new forms are claimed to be 'modern' (*xiandaide*) in opposition to 'traditional' (*chuantongde*).

Most artisans working on their own account have direct contact with many of their customers. Even those artisans working in the surrounding villages will receive the occasional visit from tourists, who might have been told that better deals can be made there than in town. As one goes up the 'hierarchy' of artisans, unmediated contact with customers becomes more common. A number of workshops have for instance been set up around Dingshu's ceramics museum and meet tourists visiting from Taiwan, Malaysia, Japan and occasionally Europe or America. As I discussed their craft with artisans working in these workshops, I understood that they learnt a great deal about the different tastes of customers from different regions or countries, and claimed to be able to tell what style of pot tourists of different nationalities were likely to buy. Masters and other reputed artisans are likely to deal only directly with their customers. This is in part a reaction to the inevitability of forgery: insisting that the purchasers visit Dingshu to buy directly from the source is the only way of guaranteeing a genuine article. Where a particularly prized item was purchased from outside Dingshu, a collector might also attempt to visit the creator of the pot to receive confirmation of the authenticity of the work.

It is precisely through this direct communication between producer and consumer that the artisan learns about the tastes of customers. Exchanges between producer and consumer are central to the economy of exclusivity that has been built around the craft. It can indeed be difficult for a customer to purchase a piece from someone they do not know. And it is part of the economy of value that artisans might decide not to sell their work, even for high prices, or might decide to limit production to only a few pieces so that these become more valuable through their rarity. One artisan had a further strategy to increase the value of his works: he refused to show anyone his work from the previous four years, and worked on only one design per year, improving it through successive attempts until he thought it was 'perfect' (*wanmei*), overall producing only a few pots in any given year for sale.

The historian and anthropologist Alan Macfarlane suggests that tea might have become one of the most popular drinks among a diversity of cultures because its preparation affords social ritual (Macfarlane and Macfarlane 2003). Unlike drinks such as beer or other brews that are prepared in one place, stored and then served at a later date, the preparation of tea lends itself more than other drinks to elaboration responding to the 'human desire to create ceremonial' (ibid.:55) since it can be prepared in the context of social gatherings. Macfarlane mentions the example of the Japanese tea ceremony, which he evaluates as 'the most extreme form of ritualization of the consumption of any food or drink in history' (ibid.:55). This ritualization relates to the sequence of steps in preparing the brew, as well as the proper handling of the different objects that are used in the ceremony. Chinese tea preparation practices are far from being as elaborate as those of the Japanese tea ceremony, though they are sometimes referred to as a 'ceremony' and often do involve prescribed practices and gestures in the sequence of tea preparation. In fact, as with so many skills in contemporary China, it is possible to pass an examination on knowledge of the tea 'ceremony', and the government-issued diploma can be an asset on one's c.v., especially useful for jobs requiring the entertaining of guests. The skilled knowledge involved in the preparation of tea is referred to by the word *gongfu*, which is also used to refer to the skills of artisans making *zisha* pots as well as a range of other crafts and martial arts. Preparing tea involves the mastering of elegant gestures in arranging the different tools and using them, and the following the proper sequence, including cleaning and heating the tea cups, and replenishing the empty cups of guests.

One might speculate that, for the same reason that Macfarlane argues tea conquered the world in part because of its potential for social ritual, in China the *zisha* teapot became favoured over other forms of ceramics for the potential for ritualization in its use and care. Indeed, the *zisha* pot itself, more than other types of vessels, also makes possible a certain number of ritualistic practices. The surface is porous and absorbs tea residues, but instead of removing the signs of use from the surface of the pot, users of *zisha* teapots encourage accretion and the development of a patina. During tea preparation, tea is poured over the pot: the first infusion of the tea leaves is not usually drunk, and is first poured into the tea cups to warm them, and then the tea from the cups is poured over the teapot. The hot tea will evaporate, leaving a residue that, through repetition of this practice, builds up on the surface of the pot over the years. Tea that has not been consumed and has become cold in the cups of guests will also be poured over the pot in this way. It is then necessary to polish the surface of the pot using a cloth; this action increases the sheen of the object, and insures a uniform patina. The handle and spout can be difficult to treat in this way as they receive less of the tea that is poured over the pot, but it is possible to use the bamboo brush that is usually part of the paraphernalia of tea preparation.

This double action of pouring tea over the pot and rubbing it with a cloth is referred to as *yang hu*, which might be translated as 'caring for the pot'. The word *yang* is also used for instance to refer to the 'raising' of children or domestic animals. The teapot is thus treated as a living object. Indeed, the action of *yang hu* creates a sheen that becomes an index of its age, thereby giving the pot a history or biography of sorts, which also impacts on the value of the object (see Kopytoff 1986). The action of caring for a pot is a tactile experience; hours can be spent rubbing it, feeling it in one's hands and turning it over to access its entire surface. Although a tactile experience, it is best to avoid directly touching the pot before it has started to develop a patina: grease from one's fingers can also penetrate the porous surface of the pot, particularly the handle since this is the part of the pot that is touched most, and this grease repels water and therefore the tea that is poured over the pot. I was told that it was thus important to handle a new pot with a cloth.

The skilled gestures that are used in caring for a pot reproduce the gestures of the artisan during the final steps of making one. The surface of the unfired pot is smoothed first with flexible tools made of buffalo horn, and in the final stage it is polished with the same kind of cloth used for *yang hu*. When I was learning the techniques of making *zisha* ware, I found it particularly difficult to master the gestures required to handle the pot whilst polishing the spout and handle in this way. This involves mastering the coordination of the left hand, which grasps the pot, and the right hand, which holds the tool; both hands move in concert to get the tool to glide over the surface. In a similar way, caring for the pot requires the acquisition of certain skilled gestures, derived from the experience of handling the object.

The practice of caring for a pot is echoed in intriguing ways in other cultural practices in China. Patina and sheen are important features in the material culture of China. Shiny things made of jade, porcelain or other precious materials have long been praised in China. *Zisha* itself is praised for its shiny surface, but unlike other items of material culture, this surface is not created by the artisan or the material used in the making of the object, but through repeated use in the context of ritualized social practices. The act of rubbing is itself of cultural significance: in temples, the bronze statues of gods or animals gain a sheen in certain areas where worshippers have repeatedly rubbed their hands over the surface for good luck. It is a common sight to see visitors in temples trace out with their fingers some particularly auspicious character inscribed on a stele or wall.

Chinese miniatures

A purchaser of a *zisha* teapot might be buying an aesthetically pleasing object and a functional tool for preparing tea, but there are also other, less patent elements that are transacted. *Zisha* teapots incorporate a range of symbols, of 'sign values' (Baudrillard *et al.* 1976), and probably most importantly, 'name values' (Moeran 1997:167). The more educated a Chinese person is, the more they are able to 'read' certain features of a teapot, and its allusions to aspects of culture. For instance, the representation of a bat in front of a doorway is an aphorism, a well-known play on the homophone *fu*, meaning both 'bat' and 'luck'. Other cultural allusions etched on teapots might be more elusive. An object might 'represent' a weight used in the past to measure rice (a truncated pyramidal shape), a traditional drum or the stone that supports the central pillar of a traditional home and so on. More than a simple game of identification of symbols, appreciation of the value of *zisha* pots can take subtle forms. The notion of 'sign' value is problematic, since the idea of the object being decoded gives only a very partial sense of the way objects are being evaluated.[1] I put forward the idea that *zisha* pots are 'miniatures' of sorts in order to go beyond the notion that the value of objects rests in their sign value.

Lévi-Strauss (1966:22–5) has argued that all art can be understood as miniatures, in the sense that every work of art has to shed a part of the

1 See Gell (1998:5–7) on problems with the idea of 'reading' art.

complexity of reality in order to become intelligible. The representation of a piece of cloth or lace in a seventeenth century painting, for instance, does not need to include every detail of it to give the impression of cloth; also, a painting, however realistic, must shed at least one dimension – volume – and both paintings and sculptures shed the dimension of time. What Lévi-Strauss identifies as aesthetic pleasure is the capacity of a work of art to reverse our understanding of the world: whereas one comprehends the external world from its parts that one fits into mental structures (the way we categorize the world), the art object 'can be grasped, assessed and apprehended at a glance' (ibid.:23): knowledge of the whole precedes knowledge of the parts.

I find the idea of art as a miniature compelling to talk about *zisha* craft. I was often puzzled when artisans would say of a teapot of simple geometric design that it 'represented' something, and would name the design accordingly, such as the aforementioned teapots in the form of a rice measure, drum or the base of a stone pillar. The act of comparing teapots to such items is deep in meaning: they refer to features of Chinese culture that are lost or perceived as disappearing. In the same way, pots that take figurative forms recall or refer to a corpus of recurring symbols in the arts of China. These include, for instance, pots representing the sun and the moon with reference to the two Chinese calendars and the *yin* and *yang*, and other plants and flowers central to Chinese painting and culture in general – bamboo, prunus, chrysanthemum, peach and so on. All these elements are 'apprehended at a glance' when someone looks at a pot and understood in terms of the image they reflect of a part of Chinese culture. In some cases, the idea of 'Chinese culture' reproduced in and on a teapot can be made extreme and might lose its subtlety; an artisan told me that among the designs she had created, her favourite was of a teapot in the shape of the map of China, with the national symbols of bamboo and the panda acting as knobs on the lid.

Chinese ancient bronzes are another source of inspiration for the creation of teapots. Given the cultural and nationalistic implications of such objects, I find it opportune to make a brief digression on this specific issue. Teapots have been made to resemble the ancient bronze ritual vessels of the Shang dynasty (1766–1123 BC) at least since the time of master Shi Dabin in the late sixteenth century. In the early 1980s, facing a high demand for new designs of teapots, some talented artisans were sent by the Zisha Factory to the Shanghai Museum to sketch Shang bronze vessels as inspiration for new creations. Although teapot designs have 'traditionally' referred to ancient vessels, I would suggest that the act of reproducing teapots in the form of bronzes has a somewhat different meaning today. Bronzes, among the most ancient cultural relics of China, bear a particular historical resonance since they are a powerful symbol of the Chinese civilization. In the context of the current ideological crisis the Chinese Communist Party is facing, the evocative power of bronzes is being incorporated in nationalist narratives on the longevity and historical continuity of Chinese civilization, with the party 'frequently invoking the "glorious past" of the nation in an effort to fill in the intellectual vacuum created by the loss of faith in the official orthodoxy' (Suresh 2002:18).

The popularity of *zisha* stems from a demand for historically and culturally 'dense' items (Weiner 1994). Such a demand can be linked to a more general trend of a revival of interest in the ancient past, and is not foreign to a diffused, although underplayed, sense of nostalgia for the 'golden age' of Chinese civilization (Hamlish 2000; Jing 2006). The subject matter artisans introduce in their work plays on nostalgia for an indefinite bygone era: teapots have always been made to resemble such objects as rice measures or stone pillar, but today the reproduction of such objects takes on other meanings since the objects being copied are no longer part of the daily lives of modern Chinese. In fact, the very existence of *zisha* today is seen, and talked about, as a remnant of an old tradition, one of the important craft traditions of China, and supposedly still utilizing the same techniques, the same designs and, probably most importantly, the same clay (the link is tangible and material) as in the Imperial era. This view is reinforced, probably paradoxically, by a discourse about the coming end of the tradition, as I explain below.

This shift to the making of objects that play on the idea of nostalgia is, I would suggest, relatively recent. With the boom in demand of the early 1980s, artisans were encouraged to create new forms, teapots that looked modern, which resulted in the most varied of forms. Today, although some artisans still build their reputation on the creation of 'modern' designs, the stress is put upon 'tradition'. For instance, one artisan, Mr Xin, has developed his reputation as the pioneer of 'modern' *zisha*. His pots resemble no others, as he purposely explores ways to contravene the conventions of the craft, thinking of ways to create pot handles that are recessed in the body of the pot, or placing lids at awkward angles. Yet listening to him talk about his work, one realizes that he puts a great deal of thought into creating objects that can be presented as culturally meaningful. His pots are sold with a written explanation, identifying a theme of Chinese culture as inspiration for the design. A pot in the shape of a fish, for instance, explicitly refers to a well-known legend, and the erotic imagery he likes to adorn his pots with is justified by the centrality of eroticism in Chinese Palaeolithic pottery. Claiming to produce traditional *zisha* pots has become even more important for many artisans since the scare of adulterated clay that was revealed by the CCTV news report in 2010, described in Chapter 1: artisans now refrain from adding (safe, I was assured) chemical compounds to create pots in a range of colours, as it has become important for artisans to be able to claim that the clay they use is 'pure', 'natural' and additive free.

In spite of the effect that a pot is meant to induce, the meanings contained within it are not accessible to everyone, precisely because some of these narratives might not be given to everyone to know. In my conversations with them, artisans often distinguished between customers who 'understand' (*dongde ren*, 'people who understand'; that is, connoisseurs) and those who do not, and they might decide not to sell to the latter or, as I mentioned earlier, sell to them at higher prices. Not everyone is able to 'see' the miniature: for this to happen, a person needs to be cultured.

Yen Yuehping suggests that *wen* (literally 'writing') and *wenhua* (culture) are central components of personhood in China, and have come to be understood as a glossing over of the self by culture:

> The character [*wen*, literally 'composition' or 'writing'] later acquired yet another layer of meaning – to paint over or to veneer over the authentic and raw material of one's given nature (zhi). Naivety is coated with layers of sophistication. What lies at the semantic core of the character wen is precisely this slow process of polishing, carving, refining, waxing and glazing of self. To be more precise, wen means to transform the natural or raw self into the social and cultural self that glistens with the gloss of accumulated heritage.
>
> (Yen 2005:46)

This transformation by *wenhua* or culture operates through education in general and, according to Yen, crucially through the practice of writing. But I would suggest that, on a smaller scale, purchasing teapots might in a similar way be a process of acquiring culture, or maybe more precisely represents the objectification of this process. Collecting *zisha* pots demonstrates that one is sophisticated, that one has taste and understands and appreciates one of the fine craft traditions of China. No doubt when a collector shows their prized items to other cultured people, they will reveal the symbolic content of a piece in the same way as the artisan revealed it at the moment of purchase.

Being able to sell a token of *wenhua* or culture means that the artisan must also be able to demonstrate an understanding of Chinese culture; they must prove that they are cultured. Many artisans I talked to mentioned the importance of knowing Chinese culture for their work. And through this same logic, many of the more successful artisans have built personal collections of antique ceramics, furniture, paintings and calligraphy scrolls in order to build *wenhua*, and most importantly a *wenhua* that can be seen and admired, and that is meant to impress. I have had occasion to observe artisans showing their personal collections to potential customers, presenting themselves as persons of taste and knowledge. The sublimation of this process appears after retirement, when many artisans decide to cover their work tables with cloth to transform them into a scholar's desk in order to practise calligraphy. It is also after retirement that artisans become most active in contributions to the published literature of the art world, a theme I return to in Chapter 7. *Wenhua* is seen as something that can be transferred to an object, as in a kind of contagious magic, to be recalled and highlighted through discourse at the time of purchase.

Following Strathern (1993), on might take the logic of the 'miniature' one step further. Talking about the art of Papua New Guinea, Strathern suggests that artworks can function as miniatures, not simply in their representation of the world as suggested Lévi-Strauss, but in their capacity to reveal social relationships; art becomes a miniature of these social relationships. *Zisha* pots also act as miniatures in this way. A valuable pot is one that is capable of revealing connections that are not only cultural but also social in origin. One reason for buying the pots of a certain artisan is that they have a connection with a master. It is evident that a large part of the popularity of Gu Jingzhou's former students is that the purchaser is in some way buying a work which contains the aura of the master. This can be understood in quite literal terms: a student's style is often very close to that of their master, a fact that is acknowledged, and usually stressed, by artisans. Also, the students of masters might have become proficient in a particular technique developed by the master, or in reproducing a particular design. Artisans cannot prevent others from reproducing one of their designs, but they will claim to be the most suited to reproduce one of their master's designs as they have repeatedly been able to observe the entire process followed by the master, and heard the master talk of some of the subtleties of the specific design. Artisans paste pictures of masters on the walls of their workshop, show visitors pictures of themselves standing side-by-side with a master, or print a family tree on certificates distributed with their pots claiming kinship with a master. Identification with famous crafts persons is central to the sale of works, and I was often surprised, and amused, at the large number of artisans who claimed Jiang Rong, one of the most important masters of the twentieth century, as either their aunt or master, with photographs and family trees to 'prove' it. When asked, artisans might admit that they received instruction from the master only for a very short time, or perhaps simply received some advice on a particular design, but as is often said in Dingshu, 'master for one day, master for life'.

In the most abstract sense, a pot can be a miniature of relations with past masters. All artisans, from the village worker to the greatest master, reproduce from time to time a design attributed to a past master. One of the most popular designs is the Gong Chun Hu – according to legend this was the first *zisha* teapot ever made, shaped to resemble the bole of a tree. In the act of reproducing such a pot is contained the claim to a lineage of artisans, a supposedly unbroken chain that extends all the way back to the fifteenth century. The very act of creation might even be said to bypass this lineage, and create a direct link with a master who lived five centuries ago. Many designs of *zisha* pots, including the Gong Chun Hu, retain the name of their original maker. In recreating a named design, artisans are creating objects that are miniatures of social relations, creating shortcuts linking the famous masters of the past with contemporary artisans competing to promote their own fame. When Moeran refers to 'name value' in his study of Japanese *mingei* pottery (Moeran 1997:167–8), he is talking of the value a pot acquires from being linked with Onta pottery in general, an individual artisan or prestigious producer household. In the case of *zisha*, name value is in the same way one of the main factors of value of the object, but in this case it extends to those associated with the artisan, either explicitly and directly (the teachers of the artisan) or implicitly and indirectly ('ancestors' of the craft).

To return to the issue of nostalgia, one might be tempted to make a parallel between the interests of consumers of *zisha* and Westerners' search for the authentic (Spooner 1986). Spooner explains Westerners' desire for the consumption of (exotic) luxury goods, in his case Oriental carpets, as driven by a need for authenticity created by our mechanized society. He defines authenticity not so much as the objective features of an object, but rather as a subjective conceptualization of the 'authentic' through the exoticization of the producers who 'still' follow traditional lifestyles. In other instances, globalization and industrialization have lead to an idealization of the pre-modern, creating niche markets for 'traditional' crafts. For instance, Susan Terrio argues that the success of hand-crafted chocolates in France is explained by the special value that such products offer: unlike their mass-produced counterparts, they evoke uninterrupted continuity with the past, which imbues them with 'cultural authenticity' (Terrio 1996:70). Likewise, Moeran places the boom in demand for the 'folk craft' *mingei* pottery of Japan within a wider context of the nostalgia of Japanese urbanites for a lifestyle closer to nature and community values in the face of industrialization and perceived 'westernisation' (Moeran 1997:140, 212).

As in the case of French chocolates and Japanese folk art, the success of *zisha* in China can also be explained by its aura of 'authenticity' and 'tradition', with strong roots in the pre-modern and pre-socialist past. What might appear more strongly in the Chinese case is that this quest for authenticity, as we shall see, is played out against a background discourse on the widespread problem of forgery and the possibilities of being deceived in the marketplace (Gillette 2010; Stafford 2006) – a discourse that is in fact not a new one, as Clunas (1991) has shown in his analysis of Ming dynasty treaties on connoisseurship. It might also be noted that the notion of the authentic might be at odds with a

Euro-American understanding of the term. As identified by Spooner, Moeran and Terrio, the desire for the authentic appears with the industrialization of Europe, and is driven by the need to recover something of the past, techniques of production that are found in other parts of the world, as if they had remained intact over time. In the case of zisha, that authenticity operates in terms of lineage: an artisan might be famous for having had a certain master, and all the more so if this master was one of the teachers recruited in 1954, who therefore represents a link with the craft as it was – in a more 'authentic' time – before the cooperative. Once again, with the question of authenticity, the central issue is one of personal relations, in this case of lineage, that connects the present-day artisan with the craft as it was in the past.

But Strathern (1993) asks another important question related to the logic of the miniature: if revelation is made possible by simplifying, what is being left out? Her answer is other social relations. For authenticity to be created, the non-authentic needs to be singled out. I elaborate on this point below, and now turn to discussing *zisha* in terms of an 'art world'.

The Yixing 'art world'

More than acting as consumers, many scholars and connoisseurs are actively involved in the creation process of *zisha* ceramics. In that sense, I would identify *zisha* craft as being part of an 'art world'. The sociologist Howard Becker, borrowing the term from art critic Arthur Danto (1964), aims at correcting the romantic idea and ideal of the independent creative artist. He defines art worlds in the following terms:

> Art worlds consist of all the people whose activities are necessary to the production of the characteristic works which that world, and perhaps others as well, define as art. Members of art worlds coordinate the activities by which work is produced by referring to a body of conventional understandings embodied in common practice and in frequently used artefacts. The same people often cooperate repeatedly, even routinely, in similar ways to produce similar works, so that we can think of an art world as an established network of cooperative links among participants.
>
> (Becker 1982:34–5)

Gallery owners, collectors, critics and others are all part of this 'network of cooperating people' and have a direct influence on the production of artworks.

Zisha craft has for long had its own art world, which has indeed had a central role in the constitution of *zisha* as a prestigious pottery tradition. In fact, the *zisha* tradition defines itself as a unique craft in China because of its close ties with the class of scholars during the Ming and Qing dynasties.

Patronage is the most obvious way in which individuals have influenced the craft. When artisans and connoisseurs associate the start of the craft with Gong Chun, it is also to stress the relationship with the literati of the Ming and Qing dynasties. K.S. Lo (1988) notes that this association with scholars sets *zisha* craft apart from the other great ceramic traditions of China, which were under the patronage of the emperor. It is interesting that Lo should write about the patronage of *zisha* pottery. He is credited (including by himself) with having 'saved' *zisha* as a prestigious craft from its fate as mere factory product when he visited Dingshu in 1979. When he writes about the singularity of the craft as resulting from its relationship with literati, he is no doubt also casting himself within the role of educated benefactor.

A similar reason seems to have driven the recent demand for the craft. The strong association of *zisha* with the literati, and their appreciation of fine teas, was at the origin of the high demand for this craft among Taiwanese, who, according to one informant, wanted to create their own tea ceremony (*chadao*) to equal the tea drinking traditions of mainland China and Japan at a time when the Taiwanese economy was booming, and with it, a new middle class searching for prestigious items of consumption. I mentioned in the previous chapter that Taiwanese merchants travelled illegally to mainland China after 1979 to acquire *zisha* pots that they could resell for high prices in Taiwan, and they also had a significant influence in the production of *zisha*.

I had occasion to meet one of these persons, Lai Jiechen, who now lives most of the year (this time, legally) in Dingshu. He recalled the low prices that were asked for pots at the time: 2 yuan per teapot made by a common artisan, and 50 yuan for a work by Gu Jingzhou, one of the leading masters – pots that are today sold for hundreds of thousands of yuan. Jie admits he did not know anything about the techniques of *zisha* when he first arrived in Dingshu. Through friendships with artisans, he gradually came to understand the process of production and learn to appreciate the craft. There were few problems interacting with artisans and managers at the factory, who welcomed the business the Taiwanese brought, and they turned a blind eye to the illegality of the transactions. In contrast to merchants from Hong Kong (who travelled legally to Dingshu after the introduction of the Open Door Policy), the Taiwanese were also in a position to teach artisans about the tastes of Taiwanese consumers. A large part of what was produced at the time, small teapots of simple shapes and particularly light, was tailored to the tastes of the Taiwanese and their tea-drinking preferences. As Jie recalled of his first visit to Dingshu, artisans then only focused on the aesthetic value of the decorations on pots; Jie gave them advice in order to improve their functionality as vessels used in the Taiwanese tea ceremony, where both the beauty and functionality of the teapot (how easy it is to use to brew tea) are essential.

There are more subtle ways than patronage by which connoisseurs and collectors influence what is produced. Today there are, for instance, a number of publications that discuss the craft. One of them is the journal published by the Yixing Ceramics Industry Association, based in Dingshu. The journal publishes articles addressing a range of topics, written by both artisans and enthusiasts. Other publications that talk of *zisha* include magazines on tea connoisseurship. After leafing through a number of volumes, I would argue that the majority of these talk about individual artisans, their life and style. Other articles talk of the history of the craft, or occasionally of special techniques. Through such publications, authors contribute to the fame of artisans, and might arguably play a role in stabilising styles, as David Carrier (2003) suggests of connoisseurs in the Western art world.

Becker sees the art world as a network of cooperating people. But tensions, misinterpretations and absurdities are also at the heart of art worlds. As Moeran shows of *mingei* pottery in Japan, these tensions result from the contrasting demands of two types of communities, to which each potter owes allegiance:

> social (the community [of potters] itself) and aesthetic (the folk art world in which their pots are distributed, sold, appreciated and discussed). The problem for potters is how best to balance their interests against the expectations of members of each of these communities.
>
> (Moeran 1997:139)

For instance, the demands of the art world call for the identification of the artisan in the sale of a work, but the idea of marking a pot with an individual's name contravenes the communitarian ideals of the craft (ibid.:164). Moeran describes artisans who have little say in the craft's art world, and are largely left out of the discussions of intellectuals who comment on and judge their work. Moeran in fact mentions that artisans decline to enter into conversations about aesthetics.

In contrast, the successful *zisha* artisan is one who participates in the art world in ways other than simply making craft objects. At the start of my research in Dingshu, I was told several times that I was wasting my time: to know about *zisha* craft, I should be going to Hong Kong or Taiwan where professors and other connoisseurs really know about the craft, yet it soon became evident that artisans knew much of the knowledge of connoisseurs and were willing to talk about the aesthetics of their craft. Much of the lore of connoisseurs trickles back to Dingshu from the discussions of scholars and connoisseurs in the specialized literature, and through exchanges between artisans and these other members of their art world. For artisans, there is a certain pride in being in both fields at once, having expertise in the making of the objects – of techniques, knowledge of the properties of the clay and so on – and being able to learn about history and aesthetics.

Some of today's masters have in this way become among the greatest connoisseurs of the craft. This is the case of Pan Chiping, who acts as a consultant to determine the authenticity of antique and contemporary *zisha* pots. Pan reads extensively about history, and learnt much from his master, Gu Jingzhou, about ancient *zisha* pottery. But for Pan, it is also his technical knowledge that helps him decide on the authenticity of a pot. He claims that this is what distinguishes him from other specialists on *zisha* craft, connoisseurs who, however much cultivated, cannot grasp some of the subtleties of the craft because they do not have access to its techniques. Becoming a connoisseur, for artisans, is seeking distinction within the community of practice of *zisha* pottery. An important implication of seeking distinction not only as an artisan but also as a connoisseur is to promote themselves as references in *zisha* connoisseurship, and acting on value. The most successful artisans learn how to talk about their craft from interaction with knowledgeable collectors; this knowledge, gained from their social networks or 'social capital' (Bourdieu 1986), can be converted into economic capital when they convince customers of their understanding and sophistication. This can be related to the development of *wenhua*, to the process of becoming cultured. One artisan had for instance learnt to write *fanti*, traditional Chinese characters, a form of writing not taught on the mainland since the 1950s but used in Taiwan, as proof of his cultivation.

In a basic sense, an art world distinguishes. Becker identifies an art world as a 'network of cooperating people', but one might comment that a network, with its implied characteristic of not being finite, must however be stopped at some point (see Strathern 1996). In fact, an art world does just that: through discourse, production patterns and purchasing practices, its members negotiate those elements that have a place in its world, and those that do not. As I mentioned in the previous chapter, in a very basic sense *zisha* needs to set itself off from the mass-produced and roughly finished pottery that abounds in the villages around Dingshu. This was done in practice, through the *hukou* system which effectively created a divide between two loci of production. It is also, and importantly, through discourses that artisans create further distinctions, setting apart certain individuals from the art world. And yet, I see village artisans as playing an important role in the art world: when they copy the work of masters, villagers democratize the exclusive craft that is *zisha*, making available the designs of masters to common people through the making of objects sold for a fraction of the original's price. One might say that there are two very different markets for *zisha*. The first is one of luxury items, made by recognized artisans and individually sold to a discerning public of connoisseurs, often successful businessmen and politicians from Taiwan, Hong Kong and increasingly from the mainland; the other market is one of lesser-value items, sold to traders by anonymous artisans to be shipped across China. Yet it is often difficult to establish exactly where mass production gives way to luxury item. For instance, artisans involved in the mass production of *zisha* pots will often try their luck in the *zisha* economy of fame, printing name cards and flyers that suggest that their work has gained recognition in the *zisha* art world. I have also heard complaints that many of the so-called 'masters' of the craft – an official title attributed at provincial and state levels – are incapable of making pots worthy of the epithet 'arts and crafts' (*yishu*).

National heritage: the *zisha* art world and the Chinese state

The nation-state plays a special role in the *zisha* art world. Responding to the 2003 UNESCO Convention for the Safeguarding of the Intangible Cultural Heritage, China has created its own national lists of intangible heritage. By 2012, China's State Council had listed 1,219 items of national intangible cultural heritage and 1,488 people as 'inheritors' of intangible cultural heritage (Mu 2012). The techniques associated with the production of *zisha* pottery were added to the national lists in 2006. China's provinces nominate items of intangible heritage for inclusion on the national lists by the Ministry of Culture. The application process, in the case of *zisha*, was undertaken by a professor at Nanjing University, in the provincial capital, in collaboration with the Yixing Ceramics Industry Association based in Dingshu. One of my informants, who participated in drafting the application, highlights that *zisha* is on the national list but has little chance of becoming recognized as an item of world heritage despite on-going petitioning. One might say that the craft has become the victim of its own success, since UNESCO's list of intangible heritage is supposed to protect heritage that is endangered, and production of *zisha* has been booming in recent years.

The younger generations seem to be more aware than their elders of the meaning of 'intangible heritage'. An artisan in his mid thirties, who had completed his apprenticeship in 2004, pointed out that applicants had selected the wrong type of ceramics for inclusion on the lists. It is not *zisha* and its techniques that are endangered, he suggests, by the more humble forms of pottery that he remembered his grandparents making, including cooking vessels, or *ong*, vessels used for preserving vegetables or brewing spirits. It is these the craft techniques and knowledge involved in making these that are really endangered, if not extinct. In the view of this artisan, the model to follow is Japan, a country that has been much more successful than China in promoting and protecting forms of intangible cultural heritage.

Others insist that craft techniques are indeed disappearing. Older artisans will often point out that the younger generation has little interest in continuing the 'real' traditional techniques, preferring to resort to plaster moulds, and are not dedicated enough to their craft in order to develop their skills to the levels of the previous generations trained at the Zisha Factory. My own observations differ in part from this reading. During my visit in 2012, I was surprised to find a renewed interest in entirely handmade pots among younger artisans. A young artisan who had recently finished his apprenticeship, Wu Jun, explained that consumers were increasingly looking for more 'traditional' pots, as opposed to striking, original designs. Although many customers who walk into his workshop cannot recognize a handmade pot, he takes time to point out tell-tale signs in order to convince them that his pots are indeed handmade. Renewed interest in fully handmade pots is also signalled by a local state-run competition established in 2010. Over the course of two days, artisans are given the task of making a handmade pot in an examination hall. A prize from this competition represents another tool to convince customers of the 'authenticity' of an artisan's handmade pots. The increasing importance of such competitions and examinations is another element that signals the presence of the state, local or provincial, and its role in defining what counts as 'traditional' *zisha* pottery.

One of my informants suggested that the reason why so many people want to learn the craft today is because of the listing of the craft as intangible cultural heritage. Intangible heritage is supposed to be 'good', but artisans, he implied, do not understand the meaning of intangible heritage. He agrees that the listing of the craft has provided much to the community of artisans: it has firstly provided great publicity for the craft; it has also promoted the documenting of craft techniques, with scholars becoming increasingly interested in researching the craft, and he named me as having played a role when, during my fieldwork, I produced videos of *zisha* techniques that, he claimed, had helped people to understand the craft.

When I asked artisans in 2012 why they thought *zisha* had become so successful in recent years, very few cited the listing of the craft as intangible cultural heritage. Artisans concurred that 2008 had been the year when demand had started to grow exponentially, and one theory was that their craft had gained international public attention during the Beijing Olympics, when the government created a campaign to promote the arts and culture of China. Whilst some artisans I talked with thought that it was a good thing that their craft had been recognized by the government, others pointed out that in subsequent years, the intangible cultural heritage lists have become less significant. One artisan joked about one of the entries on the third list of intangible heritage published in 2008: a food speciality from Zhejiang province that consists of boiling eggs in the urine of young boys, a piece of news that has been widely and dismissively reported by the international press. This was one example that devalued the significance of *zisha*'s inclusion on the lists of national intangible cultural heritage a couple of years earlier.

Zisha appears in the second list of Chinese intangible heritage. As Barmé points out in a discussion of China's heritage, the country's first two lists of intangible heritage differ markedly. He concludes:

> a closer comparison of the two lists and a careful examination of the second list ... reveals that the new enthusiasm for nominating and listing items of intangible cultural heritage might not necessarily always serve the cause of preservation. By neutralising the power and force of tradition, items of 'intangible heritage' can be the more easily commodified, exploited and commercialized.
>
> (Barmé 2006)

It is quite apparent that in the case of *zisha* the listing has a commercial purpose, rather than having to do with the preservation of craft techniques. Barmé (2005) also criticizes the State Administration of Cultural Heritage and the Ministry of Culture for having neither engaged with international debates on intangible heritage, nor developed a Chinese conception of this form of heritage. He notes that no translations of European and American key studies have appeared in Chinese, nor have many international theoretical legal studies on the related themes of cultural property, intellectual property or cultural heritage. Barmé concludes:

> Rather than define the concept of 'intangible heritage' and engage with international debate, the State Administration of Cultural Heritage and the Ministry of Culture have simply steered the debate back towards the traditional, and manageable, notion of state-legislated folk art.
>
> (ibid.)

Transactions and ties of *laiwang*

Master Gou explains that a consequence of his success, and his recognition for an iconic design of pot named the Sun and the Moon, is that, as with all other artisans in his position, he has to accept that a number of copies of his designs are being made as well as a number of outright forgeries, which even reproduce the personal name seal used to stamp his pots. Gou takes this situation rather philosophically, saying that this is his way of 'helping the poor'. But in order to respond to the situation, one of his strategies has been to insist that anyone wishing to purchase one of his works has to visit him directly at his home in Dingshu – he will not accept that individuals might act as intermediaries in his relations with his customers. Of course, this does not prevent forgery, but at least, as long as customers know of this policy, they will not trust a trader pretending to deal in Gou's works. Gou is not alone in applying such a strategy. It does of course have a drawback: artisans are in effect limiting the sale of their work. But creating scarcity is one way in which artisans affect the relative value of their work. The problem of forgery further plays a role in contributing to the exclusivity of the craft: because of the known existence of forgeries, customers might be wary when buying works other than directly from the maker.

In this context, personal connections come to play an important role. For instance, one article published in the journal of the Yixing Ceramics Industry Association is written by a collector (Li 2004) who explains that a few years back, he had wanted to establish the authenticity of a master's piece of work he had purchased from a trader. He wanted to ask the supposed maker of the pot directly, but he did not know the master personally. The author of the article explains that he finally contacted the master through a mutual acquaintance, and eventually managed to have the authenticity of the pot confirmed. This anecdote illustrates the importance for customers of establishing personal relations with artisans. Without an intermediary, the collector would not have been able to approach such an important figure.

What interests me in this article is the term *laiwang*, which the collector uses to talk of this relationship – or rather its initial absence. *Laiwang* literally means 'to come and go', and this is precisely what is expected in such interactions: the regular exchange of informal visits. Stafford (2000a:46) identifies this form of interaction as a central mechanism in Chinese culture for the building up of relatedness between those who are not, or not closely, related by kinship. Interestingly, then, the collector and author of the article implies that the ideal relationship between artisan and customer is a long-term one, that it should involve more than mere commercial transaction and should be created and nurtured through the (again ideal) repeated visits of the customer.

But what happens during the visits of customers to the private workshops or homes of artisans? I would start by suggesting that these visits are events in which different senses are involved. Most obviously, customers see many things – but they also have access to the work of the artisan through other senses, which will eventually enable the customer to get a feel for the processes of making *zisha* ware. The attention of the customer is drawn to certain elements in the artisan's house: probably before all else, artisans will show a customer certain objects in the house or workshop that might attest to their taste and knowledge of Chinese culture, such as a collection of fine antique furniture or fine examples of painting or calligraphy. Even private workshops, away from an artisan's home, will invariably have a space reserved for interactions with customers, for drinking tea and discussing purchases – spaces decorated with paintings and calligraphy, which contrast with the functional spaces of work. Serving fine teas is another way in which an artisan can demonstrate good taste – and crucially display cultural capital (Bourdieu 1986) acquired through interactions with previous tea connoisseur customers.

The workspaces themselves present a number of interesting things to see. For instance, the immaculate worktables and tools attest to the specificity of *zisha* in comparison with other ceramics traditions, since pots are slab-built from leather-hard clay, and this method is far less messy than for instance wheel-thrown porcelain. The finely crafted tools, many of which are made by artisans themselves out of fine woods and bamboo, are themselves quite unique, and artisans will refer to them to illustrate the uniqueness and refinement of their craft and its tradition. This is a point at which the sense of touch intervenes in the evaluation of the work of an artisan. Customers might be invited to handle these tools, and feel their smoothness and precision. In such instances, the tools become tokens of the perfectionism of the artisan, for whom only the most carefully crafted tools can create a quality pot. An artisan might also explain the use of these tools and some of the techniques involved, and might give a short demonstration or simply resume work for a few minutes to let the customer get a feel of how the pots are made. One artisan was able to give a thirty-minute demonstration of the main techniques in order to give an overall impression of the manufacturing process that normally would take several days. The demonstrations also involve sound – in particular, the loud rhythmical banging of the mallet against the clay during the initial stage of shaping the sheets and discs of clay that make up a pot.

Artisans constantly remind visitors that the techniques and tools employed are unique. Customers might have come across pictures in magazines illustrating these techniques, but pictures are poor substitutes for observing an artisan at work. The visits of customers to private workshops are therefore rare moments in which the demonstration of the uniqueness of making *zisha* ware can take place. The artisan must show the customer their techniques and traditional tools as a form of guarantee that their pots are indeed 'traditional'. Tradition is in this case the added value of handmade pots, and because it is not something that can actually be seen on their surface, it must be in a sense 'revealed'.

All the elements made accessible to viewing during the personal visits of customers, including demonstrations of techniques, also create a basic distinction between what is readily available for everyone to see: the open workshops of lower quality items in the streets of Dingshu and surrounding villages, and the 'exclusive' craft that is carried out in the privacy of homes and private workshops by those artisans who have chosen to continue with traditional methods. What is more, 'traditionalist' artisans claim that most of their colleagues, including recognized masters of the craft, are in fact using plaster moulds in the same way as 'common' artisans, whilst claiming themselves to make traditional pots. This revelation of traditional methods is all important: seeing their complexity, the customer might start appreciating the claims of the artisan that it can take years to master even basic techniques, and days if not weeks to craft just one pot. The value of a pot might be largely determined by the fame of its maker, but I want to argue that the high prices demanded by artisans are also importantly justified by the revelation of the skill, expertise and effort that go into making the final object – an object that, to an untrained eye, might resemble a mould-made pot that will have needed less effort and time in its realization.

Learning to see

In his discussion of the relationship between production and consumption, Dilley (2004) argues that concealing the processes of production can have as an effect the mystification of consumers. Similarly, Alfred Gell (1992) has argued that the enchanting power of art is derived from the mystery of its coming into being, which is beyond the grasp of onlookers. In the case of *zisha*, it is conversely the revelation of the techniques of the craft that transform an apparently common artefact into an enchanting work of art. Thus, in a kind of reversal of commodity fetishism, it is important for the artisan to reveal the labour that is embodied in an apparently common object. The value of a pot is derived from its links with the past through a notion of 'tradition', and from the human effort, skill and time involved in its making. These must be made visible through demonstration by the artisan. But more than simple visibility, the customer needs to get a feeling of these elements to appreciate what the artisan claims them to be: unique and traditional. There is a quality to the appreciation of these elements that must be acquired with time, and through (ideally repeated) contact with the work of artisans. As I have mentioned, this is an experience that involves other senses beyond vision: hearing, touch, the smell of the clay and even the taste of a cup of fine tea. Although the customer does not participate in the work, observing movements gives a sense of the skills – of kinaesthetics – involved in the making of the objects.

A parallel with calligraphy might be appropriate. An expert calligrapher learns to judge a piece of calligraphy by repeating in their mind's eye the movements that give rise to the characters (Yen 2005:89). In the case of *zisha*, artisans and connoisseurs alike talk of the 'style' (*fengge*) of the artisan that can be detected on the surface of their pots. This style, I was told, was the result of the development of skills: the trained movements of artisans come to acquire individuality through practice. But in contrast to calligraphy, these hand movements do not leave a 'readable' trace on the surface of the clay. Getting a sense of how a pot was made is one way in which a customer can learn to 'see' style.

I want to suggest that this learning is a kind of apprenticeship. Indeed, in many ways a customer might be said to be in the position of what Lave and Wenger (1991) have defined as 'legitimate peripheral participation' – a phrase coined as a substitute for the term 'apprenticeship', which, in the view of the authors, is too limited in scope. As they argue, novices tend to learn more through participation in social activities rather than through formal instruction. By 'being there', situated in a social setting, novices pick up techniques, but also ways of speaking or acting, that might eventually enable them to become full members of the 'communities of practice' of experts, and it is also during such participation that novices develop their 'skilled vision' (Grasseni 2004).

The position of consumers of *zisha* pots is in many ways similar to that of the novice. They too get a 'feel' for the craft through repeated exposure to experts at work, made possible through *laiwang*. The customer is thus in some regard in the position of 'legitimate peripheral participation'. Legitimate because only made possible by invitation by the artisan and mediated by common acquaintances through the principles of *laiwang*; peripheral because, naturally enough, the customer is not full participant in the craft, and will only get glimpses of its techniques. 'Participation' seems to be the only term of Lave and Wenger's concept that might not fully apply to consumers of *zisha* pottery – and in fact it has important consequences for the latter's position in the *zisha* art world, as I will discuss later.

As I mentioned, in the traditionalist artisan's view, customers' incapacity to see the differences between pots has economic consequences. There are no economic advantages in the making of handmade pots when customers fail to distinguish them from mould-made pots. Inviting customers into their private workshops, often in their homes, means that they can learn to cultivate their sense of vision. It is at least hoped that, as apprentices learn to 'see' the craft, so will customers learn to appreciate the specificities of traditional methods, and appreciate the 'authenticity' – in terms of a link with the past – of the work they are purchasing.

As I mentioned in Chapter 1, close to the end of my fieldwork, one of my main informants commented that I had become able to 'see' and understand (*kandedong*) *zisha* pots, a comment that pleased me at the time, but later brought me to reflect on what exactly this 'seeing' implied. I had gained an understanding of the technical and aesthetic principles of the craft, but I had also started to reproduce the same judgements of value about the relative quality of pots. The education of the senses is inextricably linked with power relations within the community of artisans, and between artisans and their customers: learning to see also involves learning to replicate certain values. The fact that many customers visit artisans personally has consequences here. As I have mentioned, artisans play on the scarcity of the availability of their work, and it is preferable for them to sell to people who have proved they understand the values embodied in their work – hence the strategy of one artisan I have mentioned who will ask for higher prices from people who 'don't understand' the craft.

Slowing down the flow of knowledge

I have suggested that artisans blamed the difficulties of continuing a tradition on their customers' incapacity to 'see' properly. Through this discourse, artisans were also casting themselves as the true connoisseurs of their craft. This helped ensure them a rather particular position within the craft's art world, that of superior judges of the value of their own work. In other words, at the same time as the discourse on the loss of tradition is one of powerlessness, it is also one of power, one that attempts to configure relations between the different parties in *zisha* regimes of value. Through claims to a full understanding unavailable to others, artisans are able to place themselves in such authoritative positions. The fact that artisans know more about the craft than their customers might appear common-sensical, yet that need not be the case: Brian Moeran (1997), for instance, suggests that *mingei* potters have found themselves caught up in an art world entirely engineered from the outside with the 'discovery' of this form of craft by a few intellectuals and artists who claim to discover beauty and authenticity in a folk pottery tradition. In terms of the relations of power relevant to *mingei*, Moeran tells us, artisans are not meant to understand what they are doing if they are to produce good pots, and neither are they themselves willing to comment on the aesthetics of their work. In contrast, it seems that artisans in Dingshu have been in a quite different position vis-à-vis their customers.

I have said that in a similar way to apprentices, customers learn about the craft through repeated access to workshops. But customers do not come to acquire the same familiarity with the craft. This might be obvious enough, yet this point was singled out by artisans as significant, resulting in customers failing to fully understand their work. Although everything is there to see during the visits of customers, full understanding, or full skilled vision, cannot take place. The fact that customers understand less about the craft than artisans was for the latter a great advantage, to be highlighted in discourse. Significantly, the knowledge that was unavailable to customers was the know-how of making a pot, which according to artisans impedes the full skilling of customers' vision.

I can give three illustrations of the power that is derived from artisans' fundamental capacity, and customers' failure, to 'see'. Firstly, I heard this discourse on vision during discussions with artisans about collaborations with art professors and artists. From time to time, scholars approach a skilled artisan with a design, carefully reproduced on paper, for realization in clay form. Yet at least two artisans I talked to, who collaborated in this way, pointed out that the design will never be ideal on paper. In all cases, they had to modify some elements for the finished work to correspond to their idea of a good pot. They pragmatically explained that their own work simply 'looked better' (*haokan*) than what they would have made if they had followed the original design exactly. But finding out what modification to make is dependent on actually making of the pot, it cannot be known by looking at a drawing on paper. It would in fact take them several attempts before all the 'defects' of the drawing became apparent. Because artisans have practical knowledge of the craft, they claim to be the only ones who can find out about the 'proper' forms designs should take.

The second instance concerns the problem of forgery. Often, a collector or museum representative will seek the assistance of artisans to determine the authenticity of an item in their collection. The item in question might be the work of a late master, in which case the former students of the master are deemed to be the most suited to determine the authenticity of the work. One artisan told me that he could immediately recognize, with a single glance, his former master's style (*fengge*). He was referring here to his skilled vision, shaped through intimacy with his master's work. For many years he had worked in the same workshop, and therefore necessarily, in his view, he knew more than anyone his master's works and their style. With the widespread presence of high-quality forgeries, 'looking at the style' (*kan fengge*) of pots was the only available way of determining authenticity, and collectors relied on the shared understanding that prolonged visual access to a master's work implied a discerning eye.

Thirdly, masters of the craft are often involved in the publication of authoritative works on the appreciation of *zisha* pottery. Masters have become experts not only in making, but also in talking and writing about their work. Given that books on *zisha* are one important source of information for *zisha* enthusiasts, masters of the craft exert an important influence on the way the craft is perceived and evaluated. A case in point is a volume, co-edited by three masters of the craft, that has become one of the most authoritative works on *zisha*. The book brings together the greatest antique and contemporary masterpieces of the craft, together with the works of selected contemporary artisans. That such a volume has become so important is in itself a significant illustration of the authoritative voice of masters of the craft. But even more intriguing to me is that one of the authors of this book is reported to have admitted to a friend being the forger of one of the works reproduced in the book – a pot attributed to a Qing dynasty master. I have of course no way to prove the veracity of this claim, but I find this a strong illustration of the power that artisans have carved out for themselves through a discourse asserting their exclusive understanding of their own work. If the claim is true, then the master was able to pass off one of his own works as an antique through the publication of an authoritative volume.

The latter case is further proof of the artisans' confidence that even connoisseurs of the craft are incapable of distinguishing fakes from authentic works. One might note that this also says something about artisans' valuation of the 'authentic'. In this chapter I have mentioned a couple of other cases where artisans have such attitudes towards authenticity: I have talked of artisans who pass off mould-made work as handmade, and an artisan who was relatively unconcerned about the fact that forgeries of his work were routinely being made. This I believe gives an indication as to the apparent reverence for authenticity that artisans claim to have when they talk of the disappearance of traditional techniques. One might say that, in the end, authenticity represents yet another prop that is used to sustain their authority vis-à-vis those outside the community of artisans.

Conclusion

Arjun Appadurai (1986) argued that slowing down or stopping the flow of the circulation of commodities were to be understood as acts of power in regimes of value. I have shown that artisans act to control not only the flow of commodities, by making their work difficult to acquire, but also access to full understanding of their work. Artisans act upon the way their work is perceived by their customers: they make apparent the labour and time involved in the making of *zisha* pots. This puts them in a position to stress the fundamental difference between their knowledge of the craft and that of their customers, thereby securing a position of technical and intellectual authority within the *zisha* art world. For Graeber (2001:114), it is when valuables are concealed from view that they acquire fetishistic power for future action; this concealment is often enacted through elaborate ceremonies because it must be known that objects are hidden for them to be powerful. The same with *zisha*. Artisans derive their power to act in the art world by revealing that there is something that is unavailable – concealed – even to connoisseurs of the craft.

That vision plays the central role in these negotiations of power might not be fortuitous. As Willerslev (2007) argues with reference to Merleau-Ponty, vision is particular among the senses in that it implies a distance from the object of vision. Purchasers of *zisha* pottery might be learning to 'see' the craft, but they remain impeded by the distance dictated by their position: that of spectator rather than full participant. Artisans, in contrast, are able to bridge this distance through their practice and direct engagement in the making of *zisha* pots. This enables them not only to refine their visual skills, but also to take advantage of the greater familiarity they have with the craft. 'Traditional' methods of production become valuable precisely when a craft becomes 'traditional', and by this I mean when it can be compared and contrasted with something modern, in this case plaster moulds. Traditional techniques, for their complexity, enable this discourse on seeing and understanding as they are contrasted with the simplicity (so it is argued) that a plaster mould gives to the entire process – a process that is readily available to a visitor's gaze in the less exclusive workshops that line the streets of Dingshu and the surrounding villages.

The agency of the artisan is therefore fundamentally about the drawing of visual boundaries, training the eye to see contrasts, pointing at the common to reveal the exclusive. This situation called for the introduction of vision into the discussion of the value of *zisha* craft items, since other approaches give only a partial explanation. This was related to a problem with the qualification of knowledge in economic transactions. Knowledge certainly plays a role in regimes of value, but one needs to be attentive to what kind of knowledge this is. To borrow Ryle's distinction, we are talking about a 'knowing how' as much as a 'knowing that' (Ryle 1949). Many instances of knowledge relevant for actors to situate themselves within the art world are derived from the enskilling of vision through repeated access to the work of artisans, and in the case of artisans themselves, from an intimate understanding of their intricacies.

CHAPTER 4

Learning in the factory and in the household

Mr Wan was one of my main contacts during my fieldwork. A reputed potter in his forties, he owed part of his success to having been the student and close friend of one of the greatest masters of *zisha* pottery of the twentieth century. His reputation enabled him to ask for high prices from his customers – mostly wealthy collectors from Taiwan. Wan might be crafting luxury items today, yet he started his career in the same way as most artisans of his generation, recruited as a novice worker by the original Zisha Factory. This was in 1976, and Wan at first learnt to make everyday ceramic products for export in a classroom of twenty students. Teaching apprentices in classrooms had been the first reform that the factory brought to the production of *zisha* pottery, before any technical innovation; this replaced the one-to-one apprenticeship system, with the dual aim of efficiently ensuring a new generation of skilled workers, and making the formerly secret knowledge of masters available to large numbers of individuals (see Liu and Wu 2000:114). This was of course one of the intended effects of collectivization: putting an end to the household system of production viewed as feudal and backward, and making available the knowledge of the masters for the common good.

As I explained in Chapter 1, the second half of the 1970s was a time of change, marked by the end of the Cultural Revolution and the introduction of economic reforms that saw the return of private enterprise. Renewed demand from Hong Kong – and indirectly, Taiwan – rekindled production of the kind of items that had been abandoned since the 1950s. In the 1990s, it became once more legitimate for artisans to set up work in private workshops and take on individual apprentices – in many cases their own children. Teaching in classes came to be once again gradually replaced by a one-to-one apprenticeship system, and disappeared completely a few years before the privatization of the Zisha Factory in 2002.

Wan had acquired a wealth of knowledge throughout his career, and was known as one of the few who could pass on the techniques of his master. For this reason, I was particularly interested in knowing about his own students. I knew that he had had two apprentices, but I soon realized that he was not particularly pleased with either of them: the problem was not that they were not talented, but that they were not really 'his' apprentices. The first had started to learn the craft from his own father, and only later had approached Wan to expand his repertoire of techniques. The second student had undertaken the full three-year apprenticeship with Wan. However, as with the first apprentice, the parents of the student were potters. Wan felt that both his students would not properly follow his teachings for this reason, and they would more readily take the advice of their parents.

Pressing him on this matter, I once asked Wan if he had any wish to take on apprentices under his own conditions. He confided that he had been planning to do so after his retirement. He imagined bringing together a group of about twenty students, complete novices, and leading them through the three years of apprenticeship as a group. I was particularly intrigued by this idea, in particular because Wan was thinking of recreating the learning environment of his own apprenticeship in the factory. Surprisingly, for Wan the ideal format of apprenticeship was one derived from the era of collectivization.

I had often been told that the best way of learning the craft was through a 'traditional' one-to-one apprenticeship. So why did Wan, a self-declared 'traditionalist', think otherwise? Why had the model of instruction in a factory come to represent an ideal form of teaching? Although Wan's story is atypical, I feel it raises a number of issues that deserve to be addressed. My discussions with Wan made me rethink some of the assumptions I had held about the situation in Dingshu, specifically attitudes towards learning, the advantages of exchanges between peers and the impact of the two modernizations of collectivization and the subsequent shift to private enterprise. Beyond the specific case of Dingshu, this reflection has also lead me to rethink the relationship between group formation, opportunities for learning and the transmission and circulation of knowledge.

The relationship between social participation and learning is at the core of Lave and Wenger's work on situated learning (Lave and Wenger 1991), in which the authors highlight the necessity of the social participation of novices in the community of experts in order to acquire specialist knowledge. Beyond the simple acquisition of skills, learning for Lave and Wenger comes to constitute the very identity and social position of an actor through participation in social settings, which involves becoming accepted as a member of a group through the forging of an identity that is compatible with that of the group. The use of the term 'legitimate peripheral participation' is intended to capture this double development, of 'knowledgeably skilled identities in practice and ... the reproduction and transformation of communities of practice' (ibid.:55).

Wenger has further elaborated on the concept of communities of practice, presenting it as a paradigm for organizational studies (Wenger 1998; Wenger *et al.* 2002). In a jointly authored work, communities of practice are defined as:

> groups of people who share a concern, a set of problems, or a passion about a topic, and who deepen their knowledge and expertise in this area by interacting on an ongoing basis ... Over time, they develop a unique perspective on their topic as well as a body of common knowledge, practices and approaches. They also develop personal relations and established ways of interacting. They may even develop a common sense of identity. They become a community of practice.
>
> (Wenger et al. 2002:4)

The term 'community of practice' at first glance seems entirely appropriate to describe the situation that was created in the context of the Zisha Factory in Dingshu. As I will illustrate, artisans talked to me about the sense of solidarity that was created in the context of learning the craft in classrooms, which, together with Chinese notions of social obligations, combined to uphold the ethos of sharing that was established during collectivization This in turn enabled the exchange of information and knowledge about the craft: techniques, designs, knowledge of the demand for certain kinds of pots and avenues for their sale. Wan was explicit here: his students would be able to help each other when a lesson had not been understood, and a sense of solidarity would develop among them. But Wan's plans had also been formulated in terms of exclusion: he perceived the fact that his previous apprentices were under the influence of other individuals, their own parents, as a problem, and he wanted to address this by creating an 'exclusive' group made up of novices – who would not have to 'unlearn' bad habits – and a situation in which he would be the sole expert.

The notions of exclusion and secrecy – relating to specific techniques or to spells or rituals – and more broadly, the agency of artisans in controlling the circulation and dissemination of craft-related knowledge, are at the centre of many anthropological accounts of craft. Marchand (2001), for instance, comments at length on the difficulties builders of minarets in Yemen have to overcome to be accepted as apprentices by masters in order to learn their secrets. Herzfeld (1995, 2004) talks of the analogy of stealing the knowledge of the master in the context of apprenticeship in crafts and petty trades in Crete, whilst adding that masters themselves encourage such attitudes to test the mettle of their apprentices. Theft in these and other contexts is what defines the boundaries between insiders and outsiders (Herzfeld 1995:131–2). Exclusion can also be operated through discourse: Kondo (1990:229–57) reports on the narratives of holders of expert knowledge in Japan, who through narrative cast their role as holders of traditional knowledge and personifications of 'Japaneseness'. The effect of these narratives is one of exclusion, of women and younger workers who are not included in the definition of holders of expert knowledge. For Wenger *et al.* (2002:139–40), exclusion is indeed a defining feature of communities of practice.

How can solidarity and exclusion coexist within communities of practice? Wan was trying to recreate the solidarity that he experienced with his former classmates and colleagues, whilst at the same time seeking exclusion. Wenger, Herzfeld and Kondo talk of exclusion from communities of practice as necessarily implying exclusion from social interaction. This must certainly be true, but it does not account for an alternative situation presented by my case study, where, as I will explain in more detail, individuals resort to certain tactics to both uphold social interaction whilst controlling the flow of practical knowledge.

In this chapter, I follow the paths of elements of design and techniques as they are passed on from master to apprentice and between peers, and the paths of sociality along which artisans move with the dual aim of exchanging gossip and learning about new techniques or designs. This will enable me to reveal some discrepancies that exist between the 'community' shaped by social interaction and the 'community of practice' originating in the Zisha Factory.

From *yijia yihu* to the Zisha Factory

In the minds of artisans in Dingshu, there was clearly a 'before', 'during' and 'after' the Zisha Factory in terms of how apprenticeship was conducted. Through interviews, I tried to reconstruct the features of apprenticeship before 1949, and in the factory from 1954 until just after the economic reforms, using these periods as a basis for comparison with the contemporary situation. The expression used to describe the mode of *zisha* production before the time of the cooperative was *yijia yihu*, which can be translated as 'every household'. The term *jia* in Chinese can refer to many things: the physical structure of the house, the household, the family and an economic unit. In Dingshu, *jia* implied all these meanings. In this system, each household was an economic unit, and before collectivization, households specialized in a specific style of pot, while children learnt from their parents to carry on the business.

This was the case for Jiang Rong, born in 1919, who is considered one of the greatest masters of the twentieth century and who joined the Zisha Workshop in 1955 as the only female master. She came from a family of artisans, and learnt the craft from her father from the age of 11 after she dropped out of school. Later, when she was twenty, she studied with her paternal uncle when she joined him in Shanghai to create reproductions of antiques. This seems to be a classic example of apprenticeship within the family, whereby a child learns the style of a family member.

When teaching took place within a household, however, it did not always mean that a parent taught their child. Gu Jingzhou, for instance, who also joined the cooperative as a teacher, did not learn from his parents. The latter were not artisans themselves, but *yaohu*, kiln owners, who owned land and employed artisans who worked on *zisha*. Gu claimed to have learnt from his paternal grandmother in the early 1930s. This, however, might be only partly true; a student of his – and one of my informants in Dingshu – maintained that he most probably learnt from one of the artisans employed by his father, who would have been paid to be a tutor. The reason why Gu was unwilling to acknowledge this artisan as his master is that the relationship had not been formalized, and Gu had not paid respects to the master (*bai shifu*), an act that formally marks the beginning of the relationship. Coming from a relatively wealthy family, Gu had received a formal education, and only started learning pottery at the age of eighteen, at a time when his parents encountered financial difficulties. Eighteen was a relatively late age since it was normal at that time for apprentices to start at the age of fourteen or fifteen.

Another case illustrates the fate of less wealthy families. Children would be sent away to be apprenticed with the hope that they would bring in a new source of income. This was the case of Wang Yinchun, who also became one of the first teachers at the Zisha Workshop. Wang came from a poor family unconnected with the pottery business, and was sent to learn potting in the house of a master. For three years he lived in his teacher's home, his main task being to carry out household chores, and only occasionally did he learn some basics of the craft. Wang came out of this apprenticeship period with very little knowledge of the craft.

Because of the economic situation at the time, few individuals could afford to carry out their apprenticeship in the 1940s, a situation that remained until the creation of the cooperative in 1954. Xu Hantang, a present-day master, recalls how he started learning with master Gu Jingzhou in the late 1940s. Gu was a friend of his father, who ran a business and sold Gu's work, and Gu took on Xu as an apprentice out of a feeling of obligation to Xu's father. Teaching used to take place in Gu's house, on an irregular basis since Xu had to cover several kilometres on foot to reach his teacher's house. The apprenticeship did not last very long, however. Due to the difficult economic situation, Xu and his teacher were soon forced to leave for Shanghai to find work, and it was not before the start of the Zisha Workshop in 1954 that his apprenticeship resumed.

The last case is rather interesting to me, since it shows that the choice of master or apprentice could operate along the lines of friendship, or more generally, of obligation. As I will show, the present-day situation presents many similar cases. I would argue that *yijia yihu*, more than describing a reality of the past, was talked about as an ideal scenario to underline the importance of the household as the locus of production, and to set it up in contrast to the situation(s) that followed.

The creation of the Zisha Workshop had several important consequences for the process of learning. Apprentices were recruited in the same way as workers; family ties or social obligations no longer interfered with the choice of students. Master Wang Yinxian recalled how she had joined the factory after passing an entrance exam aimed at demonstrating her capacity to work with their hands, which involved cutting square and round clay forms This also meant that there was no possibility for masters to choose their students, nor the reverse. Because they were recruited with a view to becoming fully employed by the factory, students had far more time to concentrate on learning than in the past, when they might have been asked to carry out chores, as with the example of Wang Yingchun that I referred to above. The aim was to get apprentices to be fully productive by the end of the term of their apprenticeship. Although the apprenticeship of early students was to last three years, that term was cut short in 1958 when the first apprentices became the tutors of hundreds of new recruits.

Another revolution involved teaching in classes of ten to twenty students, replacing the traditional one-to-one apprenticeship. Teaching in large classes had, I would suggest, two important effects. The first was that teachers had to adopt certain teaching methods, and to articulate teaching through the developing of theory – in other words, becoming more 'formal'. Most patently, Gu Jingzhou developed a 'curriculum' consisting of eighteen lessons that addressed all aspects of the making of *zisha* pots, from design and aesthetics to tricks of the trade. In these lessons, he introduced elements of mathematics to talk of the proper way to design and make pots. For instance, he taught that the line of the perfect pot must be that of a parabola (*paowuxian*). In a photograph of the time, one can see one master sitting behind what looks very much like a school teacher's desk, and behind him a blackboard with the design of a teapot drawn in section view.

The second effect of teaching a large class was that apprentices could rely on classmates for help. Teachers would explain the action to be performed, and would retire to their own work until ready to check on their students' progress. If a lesson had not been properly understood, students could rely on classmates for help. This also introduced varying degrees of solidarity and competition within the class. This is how one present-day master, Wang Yinxian, described how she joined the factory, and her period of apprenticeship there:

> When I was young there was a playground near school. In 1954, there was a big exhibition in the playground, and many products of pottery were shown there. Fortunately, I visited it, and saw all kinds of little teapots. I thought these were very beautiful. I was deeply attracted, because my father worked on pottery, but after I saw such beautiful teapots I was moved and I thought maybe it was a good idea to work on pottery. This was my idea after my visit. In 1956, because I had many sisters and brothers, my family was poor. Just at that time, the Zisha Workshop was looking for a second set of students. When I learnt the news I decided to join. After some simple tests ... I was accepted. In the workshop, we learnt techniques from the masters in the daytime, and in the evenings we studied culture ... At that time the students did their best ... We had to learn the basics, such as making clay discs and sheets. Every step had to be tested. The teaching style was very good. We were brotherly and sisterly, and we studied hard. On Sundays we went to the classroom to practice. In the work room, the conditions then were very bad. There were no electric lights, we had to buy our own candles. The competition among students was fierce. Whatever someone did I had to do better. However, still now I miss that period.[1]

The sense of community evoked in this account represents, I would suggest, the greatest 'modernization' of the craft, far more than making or firing techniques. The fact of having been taught in a classroom, of having classmates, represented a significant change, and created strong bonds between people – although the evocation of competition is also interesting to note in this account. I will return to these themes further on in this chapter.

In the early 1980s things started to change, following not only the economic and political transformations of the country, but also the renewed interest in *zisha* pottery in Hong Kong and Taiwan, as mentioned in Chapter 2. With the demand for fine, original and, most importantly, named works (the practice of signing pieces, interrupted during the Cultural Revolution, had resumed in the late 1970s), the Zisha Factory created a research department (*yanjiusuo*). This involved creating individual workshops, separated from the communal ones by walls, in which the most talented artisans had the freedom to work on new designs and technical innovations. Some artisans just out of their three-years apprenticeships were recruited by these masters as helpers.

1 Personal communication, 3 November 2003.

They continued learning with them, and to some extent the practice of one-to-one teaching resumed in this context.

The economic reforms also saw the reappearance of the family as a potential economic unit (see Bruun 1988; Cohen 1992), an ideal context in which artisans could teach the craft to their children. The reintroduction of working at home meant that the children of potters would be in immediate contact with the craft. Because apprenticeship is a lengthy process, requiring intensive supervision, this seemed to have be a logical choice, and many among the new generation of artisans learnt the craft at home. This happened in two different stages. In the 1980s, artisans produced pots illegally at home at night, at a time when the two existing *zisha* factories (the original Number One Factory, and Number Two Factory, established in 1980) could barely keep up with demand from Taiwan. In that context, when artisans resumed production at home, the craft once more became visible to children on a daily basis. Then, starting in 1995, when the Number One Factory stopped providing a guaranteed income to its workers, the latter worked for themselves, either in their workshop in the buildings of the factory or at home, this time legitimately, registered as *geti hu*, or independent producer.

Mr Fang, who himself taught his son at home, suggested that the situation had gone full circle in the space of a few decades. In his view, after the cooperative factory's attempts at rationalizing and democratizing teaching structures, the present-day situation shares features with the one that existed before collectivization. Fang however did not imply the resuming of the *yijia yihu* system, and as a corollary, the closure of knowledge that is seen as characterizing that era. I rather see the contemporary situation as something of a hybrid between the apprenticeship systems of the collective era and earlier times. In order to explain this, I want to start by asking whether teaching within the family is really an ideal, and consider some cases in which it is preferable for a child of an artisan not to take up the craft.

Kinship, talent and friendship

There can be advantages for an apprenticeship to occur outside the immediate family. Goody (1982b), for instance, reports that in Ghana it is preferred that apprenticeships take place not with parents but with more distant relatives, or even outsiders living in town. This reason for this is twofold: learning different sets of skills that the parents might not possess or fully master, and because it is feared that parents might be too lenient with their child's education. Singleton, talking about *mingei* pottery, mentions that in Japan, masters are rarely kin; as the saying goes, 'you have to eat someone else's rice' (Singleton 1989:22). Although apprentices are integrated into the family, with absolute dependence on master and family, and share the meals and patterns of life of the family, this is not the dependence a person has on natural kin; rather, it is linked (explicitly) to that of a wife moving into a household, who must learn the habits of the family, and deliberately train for dependency. The master never compliments an apprentice, in the same way as a father-in-law never compliments a daughter-in-law. Given the strong patrilineal bent of Chinese society, one might expect that the passing on of knowledge from father to son might be the norm, and indeed it is tempting to think of patrilineal kinship as the model for apprenticeship. DeGlopper (1995:25) argues, however, that relationships between masters and apprentices are not modelled on kinship in Chinese culture.

In Dingshu, in many ways it is desirable that an apprentice be the son or daughter of an artisan. A major advantage for the apprentice is that the fame of a parent can greatly help an apprentice's career. As already mentioned, genealogical trees are often reproduced on pamphlets presenting an artisan's work. For some families, this lineage can reach back to the period before the family system was dismantled in the early 1950s. This is the case of the Yao family, whose members are proud to say their potting business is five generations old, and who in a way have recreated the *yijia yihu* of old, transforming their family name into a brand name. Artisans from the family sell objects that are priced according to seniority: the two most senior artisans, who were trained in the early years of the Zisha Workshop, sell some of the most expensive works in Dingshu. If a customer cannot afford the work of the masters, they are invited to buy the work of one of the sons, who has also gained a strong reputation and sells his work for relatively high prices, or if this is still too expensive, the latter's daughter, who finished her apprenticeship a few years ago and before that graduated from an arts school, sells her work for less. The pots made by a nephew who is close to finishing his apprenticeship with the grand-daughter's husband are not yet sold but regularly given away as gifts by the more senior artisans. Buying the work of a younger artisan in the family might not guarantee a piece of high-quality workmanship, but there is a guarantee (or a claim) that the work uses clay of good quality, that it is made using traditional methods and that it has received the approval of the young artisan's seniors. This example points to ways in which some of the advantages of the old family system can be recreated.

Sometimes, however, it is neither expected nor desired that a child carries on the work of the parents. I often asked parents with young children whether they would like their child to become a potter, and the usual answer I got was that it was up to the child to decide. Although parents can be very demanding with regard to their children's success, there is still the notion that the child has a degree of freedom in choosing their path. Indeed, because many artisans have been very successful, they can often afford to send their child to university. I was also often told that the future of craft ceramics is uncertain, artisans being only too aware of their dependence on a market that is susceptible to changes in fashion. The *zisha* teapot is a status symbol that can easily be replaced by another; artisans got a taste of that reality with the crash of the Taiwanese economy, when overnight their businessmen customers lost interest in *zisha* pottery. Given such instability, it is understandable that parents might prefer to see their child turn to a different, more economically sound, activity. The children of artisans are then just as likely to be bankers, teachers or factory managers than successful artisans themselves. Interestingly, however, during my return visit in 2012, many of those younger persons had returned to Dingshu after being unsuccessful in finding appropriate or sufficiently remunerated work in the cities in order to learn pottery; with the boom in demand for the craft, and because of the advantages, including monetary ones, that could be had in living with or near parents (for instance, help with childrearing), being an artisan was economically more desirable than many white-collar jobs in the cities.

Another obvious reason for a child not following in their parents' footsteps is simply the absence of talent. I was often surprised at how discouraging some parents could be about their child taking up pottery. For instance, a man in his twenties who was the son of a couple of artisans was studying at a university in a nearby city and did not seem the least interested in the work of his parents. But I soon came to realize that he had been discouraged from taking up the craft in the first place. His father often talked disparagingly of his son, telling me once he thought he was childish. I understood that an artisan's talent should not necessarily reflect on their child, and this understanding is in essence what makes some artisans look for apprentices beyond the family circle. Yet the economy has an important role to play: when I returned to Dingshu in 2012, this young man had in fact returned to Dingshu after graduation to start an apprenticeship. As have many young people, he had discovered that a university degree did not necessarily lead to a highly paid job, especially in the new, difficult economic climate; production of *zisha* had become a more stable and promising activity than previously assumed.

The discourse on the presence or absence of talent rather surprised me. I had expected, from the anthropological literature on skills and apprenticeship, that talent would not come into account. For instance, Singleton (1989:22), speaking of *mingei* pottery in Japan, explains that the success of an artisan is only explained by hard work and discipline, and not talent. With a craft like *zisha*, in certain respects even more rigorous than Japanese Mingei in its precision, I had expected a similar understanding of hard work. But in Dingshu, there is a discourse surrounding the presence or absence of talent. The term *wuxing*, which I translate as 'talent', first came up in a discussion on the apprenticeship system in the Zisha Factory.[2] I was told that teachers would early on decide which of their students had *wuxing*, and which did not. This had as a result the promotion of the talented to the 'advanced' classes, and in some cases the interruption of the apprenticeship of the non-talented – the student would be placed in another, less demanding, job in the factory.

The first year of an apprenticeship involved the learning of the basic stages of the craft, the beating of clay into discs and sheets, and was crucial in determining which student had *wuxing*. The movements that are learnt during this first year are feats of coordination: the making of clay discs involves using a mallet held in the right hand to pound the clay with a vertical movement, whilst the left hand must, in between blows of the mallet, pick up the clay disc from the table and turn it a few degrees. I see this 'talent' determined early on in the learning process as in large part defined in terms of the coordination of movements.

The importance of interrupting an apprenticeship is also explained by a crucial element of the techniques of *zisha*, of which I will talk more extensively in the next chapter: the importance of sequence. In the making of a pot, the same care must be put into all the stages, including the initial and basic ones, because it is not possible to rectify a mistake later on. If the clay sheets are not perfectly flat when first made, then at no subsequent moment can that fault be corrected. Artisans referred to this principle of the craft whilst commenting on apprenticeship: an apprentice who is not doing a good job at an early stage is better off trying another activity. Some individuals I met would tell me that they had tried learning the craft but had stopped after a few months, and they would explain that they where not suited to the job. One woman used the expression 'I do not eat of this rice', which might be glossed as 'this is not my job'. An artisan's career, at the time of the Zisha Factory, was then decided at a very early stage of apprenticeship. The situation is different in the post-factory context. It might not be in the interest of the teacher to interrupt an apprenticeship, notably when the apprentice is the son or daughter of a friend, as in the case I started this chapter with. The notion of *wuxing* remains, however, key to evaluating the worth of students, and parents might look for signs of talent in a child and either encourage or discourage a career in the craft.

I started this chapter with a case that presents another situation where the 'ideal' of a parent teaching their child seems to be negated. Wan's two students

2 The Chinese-English Dictionary of the Beijing Foreign Studies University translates *wuxing* as 'power of understanding; comprehension' (*Han-ying cidian*, Beijing Waiguoyu Daxue, 2003 s.v.). In my case, I prefer the word 'talent', which I think is justified in the sense that, as it was explained to me, it is understood that some people have *wuxing* and others do not; it is something people are born with.

were the sons of former colleagues from the factory. The reason why they had been sent to Wan to learn the craft was explained by the fact that, although the parents were artisans themselves, they thought Wan was more advanced in certain techniques. Left unsaid, I believe, was the fact that an apprenticeship with Wan was a way of 'adding a name' to the apprentice's c.v., in addition to the name of the parent. In many cases, when I asked apprentices how they had found their master, or masters how they knew their student, the answer would refer to the fact that the father or mother of the student was a friend (*pengyou*) of the master. This rather surprised me at first, since friendship is said to be rarely recalled in China to explain a relationship (see Smart 1999). Referring to the relationship as one of friendship indicated that the student was not paying for their tuition. It was indeed presented as if the teacher was doing the apprentice, and the latter's parent(s), a favour. Pressing on with my questions, I sometimes got to understand in what way a father and a master would be friends. In Wan's case, the father of one student had had the same master as Wan, and the parents of the other were colleagues from the Zisha Factory – and all were Wan's neighbours. Of the apprentices studying in the workshops of the factory, three of them explained that they had found their teacher through a tie of friendship – in two cases through their father, and one artisan through her husband. In the latter case, for instance, the teacher had collaborated with the apprentice's husband, who engraved the teacher's pots with calligraphy and pictures.

I feel there is still a need to explain further the idea that 'friends' are suited to teach one's son or daughter. I would argue that in large part this derives precisely from the situation that was created by the Zisha Factory. I am interested here in the 'paths' knowledge is taking. In what directions is transmission operating? What is the significance of knowledge travelling along lines of 'friendship', and not, significantly, descent? To understand this, I think it important to shift perspective from apprenticeship to exchanges between peers. This will help us both to understand choosing a master, and to conceptualize these 'paths' of knowledge.

Exchanges between peers and the notion of *lai baixiang*

Artisans, as they say themselves, 'never stop learning' throughout their careers, and it is through exchanges with fellow artisans that new designs, tricks and techniques can be learnt. As I became interested in these exchanges between peers, I realized that in fact much of the information that artisans learn is not transmitted during apprenticeship but later on in an artisan's career. I have mentioned that the doors of workshops in the building of the former factory are usually open, and people exercise the right to walk in uninvited for a chat. Usually, fellow artisans visiting a workshop will take a moment to look at the work in progress with an expert eye (at times feigned) and offer unsolicited advice (often ignored). I was at first a little surprised, and amused, by these attitudes, and particularly by the usual absence of greetings and farewells. It reminded me, however, of Stafford's (2000a) writing about the cycle of *laiwang*, or reciprocity between non-kin. *Laiwang* is a term used in everyday speech, literally meaning 'come and go'. One can say to have *laiwang* with someone, meaning that one is familiar with that person. This implies that one is expected to invite the person to all the significant ceremonies of one's family, and to be invited in return. But also, more generally, people engaged in relations of *laiwang* should pay regular visits to each other.

Stafford highlights the importance of arrival and departure, and the rituals associated with these moments, in many aspects of Chinese culture. He argues that the absence of greetings is an indication that, between close kin or friends, one is never really separated, and the idea of greetings or farewell would negate this ideal: 'because emotional and/or practical separation *should* be impossible between those who are genuinely close, and genuinely engaged in each other's lives, a mere physical departure is, in this respect, inconsequential' (ibid.:66).

I only rarely heard the term *laiwang* in exchanges between artisans. Instead, people usually called others, asking them to 'come to spend some free time' – my translation of *lai baixiang*, a Wu dialect term (here rendered in Mandarin Pinyin), and the equivalent of the Mandarin *lai wanr*, 'come to play'. This referred to 'passing time' with others, such as playing cards, mahjong, or most of the time simply chatting and gossiping. *Lai baixiang* is related to *laiwang*, in the idea of regular visits people should pay to others, as well as crucially, movements of coming and going – and the absence of greetings that seemed to negate the idea of separation. The movement of *lai baixiang* is to me the essential context of the flow of knowledge that takes place between peers. I would argue that this flow only takes place within existing relationships of reciprocity, or at least needs the creation of such ties in order to take place.

These unsolicited visits to the workshops in the old factory's buildings usually took the form of a casual chat, but even then ideas about techniques or aesthetics were often exchanged. Most simply, anyone visiting a workshop can watch work in progress (and the artisan there might continue working while chatting), or see the person's latest work on display, and draw ideas from that. These are importantly occasions when the work of the artisan is evaluated by others. A fellow artisan walking into a colleague's workshop will usually start by commenting on the work in progress: *'Hao kan!'* or *'Tai pang'* ('This looks good!' or 'It's too fat', meaning the pot looks too plump), and a discussion might arise in response about what the proper shape of the pot should be. At times, visitors would ask for more precise information. Mr Yi is a respected artisan, and some individuals would visit him from time to time to ask him for advice, at times about specific techniques. I saw him giving short demonstrations, for instance how to make a particular type of spout, or giving advice on which type of spout would be most appropriate for a type of body. One main advantage of seeking advice from Yi rather than from another master was that his workshop was in the building of the former factory, and in contrast to a private home, people could visit at any time, without having to knock on the door (though Yi did occasionally close and lock the door for a moment of peace).

Working at home might have the great advantage of privacy, but it might also become a limitation insofar as one risks loosing contact with others when working away from the old factory. For this reason, artisans working at home often pay more or less regular visits to those former colleagues who have found it easier and preferable to stay at the factory. Indeed, even though the factory is no longer the central institution it used to be, it is still the main place where people meet. Besides a couple of public parks, expensive teashops and restaurants, there are no real alternative public meeting places in Dingshu where artisans could otherwise meet and exchange ideas.

Wan was one of those artisans who at least once a week visited his colleagues at the factory. I was at first rather intrigued by these comings and goings, and he was a little elusive about what these visits implied. Eventually, I understood that in addition to an occasion for a chat with friends, he would offer advice and a teacher's thoughts to others. But he did not just visit any workshop. Some people seemed to be on his 'list'. One artisan, Mr Hu, would tell me proudly that every time Wan came to the factory, he paid a visit to him. This was significant for him, a way to stress the tie he had with a very famous and talented artisan. It is interesting to point out the basis of this tie for what follows: Hu and Wan had been classmates during their first year as apprentices at the Zisha Factory in the mid 1970s. The Yao family I discussed above might have recreated a semblance of family production, but they too tapped into the networks that originated in the factory – for instance the son-in-law mentioned was the apprentice of an unrelated potter who had worked, like his father, in the factory. Once an artisan working at the factory talked to me in rather harsh terms about a colleague who no longer participated in social activities and did not even visit people at the factory any more. He further talked of the work of the artisan as having been rather good at one time, but having become poor in quality in recent years. The link was not explicit, but he seemed to suggest that cutting one's ties with the social life of the old factory implied an artisan's inability to develop their craft.

Sociality and learning are the two indistinguishable aspects of *lai baixiang*, as the comings and goings of sociality melt into occasions for learning (and teaching). Discussions might start with evaluation of the work in progress, would switch to gossip, and maybe to a quick demonstration of a principle, or leafing through a book on *zisha* craft to prove a point. Conversely, the absence of sociality hinders self-improvement.

I would thus suggest that knowledge moves along the paths of sociality created by *lai baixiang*. Stafford (2000b:105) mentions that the notion of *laiwang* is related to the notion of *guanxi*, and indeed, an understanding of *guanxi* can explain further these exchanges created through *lai baixiang*. *Guanxi* is often presented in the literature as the defining feature of Chinese society.[3] Relations of *guanxi* (loosely translated as 'connection') are personal networks, dyadic social connections that are long term, and are called upon in exchanges of favours. These relationships tend to be built upon shared identities such as native place, kinship or attending the same school (King 1991; Smart 1999; Yang 1994). Some have seen *guanxi* as primarily instrumental, but others have stressed the essential dimension of affectivity. Smart (1999:129) argues that *guanxi* derives its efficiency from the combination of instrumentality and solidarity.

It is interesting to note that this fostering of ties between artisans in the *zisha* community of practice seems to contrast with the situation in Jingdezhen, another major ceramics production centre in China, known as the 'Capital of Porcelain.' Maris Gillette (2010) describes the intense competition that exists in Jingdezhen after the turn to capitalism, and mutual mistrust among artisans, who fear that competitors might steal their secrets. In this context, Jingdezhen's artisans have opted to rely extensively on ties of kinship and common origins (persons coming from the same village) in forming business partnerships or employing assistants. Rather than debate the relative importance of the family in capitalist contexts, I believe it is important to recall that relatedness in China (as elsewhere) is constructed (through everyday exchanges of gifts, visits, caring and favours) rather than 'given' (Santos 2008; Stafford 2000a). As has been repeatedly noted in the anthropological literature on China, establishing more or less instrumental social relations (*laiwang* or *guanxi*) depends upon constructed claims to shared identities, for instance having attended the same school or sharing the same birthplace (Gold *et al.* 2002; Kipnis 1997; Smart 1993; Yang 1994). Susanne Brandtstädter (2004), discussing economic learning in a village in China's Fujian province, argues that the turn to capitalism did not mean the 'unlearning' of the practices that produced community. In Dingshu too, artisans have not given up the values of community and exchange that reproduce it: the turn to private enterprise did not mean the dismantling or

3 See esp. Yang (1994). However, the debate is open as to whether it is a specifically Chinese institution or the Chinese version of a common phenomenon: see Gold *et al.* (2002).

negation of community values. On the contrary, their continued success as ceramicists depends on the maintenance of such values.

Seldom discussed in the literature on social relations in China is the way new forms of relatedness might have been created by collectivization. Artisans in Dingshu stressed that the collectivization era created a new kind of community, and provided new opportunities for making claims to a shared identity. Most interestingly, the factory made possible new kinds of claims of relatedness, notably as 'classmates'. Although such relationships existed in the context of formal education, they did not exist in the context of craft apprenticeships in the past. In the present capitalist context, relations between fellow apprentices and co-workers from the factory provide the chief 'common identity' that artisans call upon in daily exchanges.

In Dingshu, I have found that it was crucially relations that had been created in the context of the Zisha Factory that were called upon in exchanges of expertise, and it is these relations that might be called upon to find a suitable master for one's child. When addressing the notion of *guanxi*, many have seen it as a practice that provided solutions in the context of the Cultural Revolution, where one had to rely on social connections to access certain goods and services that the state system failed to provide (e.g. Gold 1985; Yang 1994). Interestingly, in my case, the situation is somewhat reversed. I found that artisans rely precisely on the ties created by a former state institution as the basis for exchanges of favours. As I mentioned, narratives about the creation of the Zisha Workshop would stress the break in the closure of knowledge (which characterized the *yijia yihu* system) that came about when masters and a generation of new students were regrouped under the same roof. That communal era has become an idealized moment when social ties were created seemingly out of nothing. Whereas in the past, a master taught one, sometimes two, students at a time, after 1954 every apprentice had twenty classmates; classmates helped each other, gave support to each other, and later, even after the end of the Zisha Factory, these ties remained strong. It is through the present exploitation of these ties that artisans can perpetuate some of the services and advantages that the now privatized factory no longer provides, in particular teaching novices and access to knowledge.

Once an apprenticeship is agreed upon, a new social relationship is established between apprentice and master. Artisans would often repeat to me a saying, 'once a teacher, always a teacher', to characterize the nature of this relationship. This referred to the expectation that the student should act in an appropriate way towards the master, even if the actual period of teaching had been short. In the first year of my fieldwork, a prominent master died, and one of my close contacts was among those, including the family of the deceased, who mourned the master. I was told that this particular artisan was expected to mourn the master given that he had been her student for one year, the first year of his apprenticeship. The phrase 'once a teacher, always a teacher' was used in another way by artisans: when asked who they had learnt the craft from, most would refer to the name of a well-known figure in the community, even if, after further enquiry, the artisan was found to have learnt from that particular master for only a short period of time, or simply received comments and advice. In answer to my question 'Who did you learn from?' I would thus most of the time be given the name not of the person from whom the artisan acquired most of their knowledge, but the most prominent figure among all those whom they had asked for advice. This might be interpreted as a form of deception, and indeed many artisans play on this to promote their work: a photograph of them posing with a famous master is often used to claim having been the student of that master. Still, beyond these claims, it is expected that the relationship between master and student is durable in time. A student should express *renqing*, or 'human feelings' (see Yan 1996:122–46), towards a master. This is most often expressed in the form of gifts – of alcohol, cigarettes or tea – on the occasion of important Chinese festivals such as the New Year. This expression of feelings in the form of gifts has a non-negligible monetary value, given that alcohol and cigarettes are expensive. One informant, a young artisan in his late twenties, explained that he had never paid any of the several teachers he had learnt from. This would be the norm in Dingshu, related to the fact that it is usually relatives or family friends who are approached to become the masters of aspiring young people, and monetary payments should not be part of such relationships. But the young artisan added that the regular gifts of alcohol and cigarettes to his former masters were understood by both parties as representing some form of monetary compensation.

Concealing and stealing knowledge

In spite of the central place of sociality between former colleagues and friends in the transmission of knowledge, there might be times when artisans might decide to reduce opportunities not only for exchange but for claims to certain forms of knowledge. Artisans do so by participating in a selective manner in the movement of *lai baixiang*. I did not find, in Dingshu, an aspect of apprenticeship that has been described by other anthropologists, the idea that for the apprentice to learn, they must steal knowledge from the master (e.g. Herzfeld 1995, 2004; Kondo 1990:238; Singleton 1989:26–7). I would suggest that instead there is an ethos, created by the Zisha Factory, that the basic knowledge of the craft should be made available to any apprentice. Artisans go about visiting their colleagues in other workshops, and might offer their advice to apprentices learning there. When I was learning the craft in Zhao's workshop, I had no less than four regular 'teachers' besides my 'official' teacher, who would come in for a chat, but kept an eye on what I was doing so that they might correct a mistake.

The openness of knowledge derives from the situation that was created by the factory, where knowledge became freely accessible to everyone working there. However, I do not think of this knowledge as becoming 'common property', any more than there was a notion of private property or ownership before the factory. In fact, there is no notion of knowledge as property in Dingshu; designs are not said to be exclusive to one artisan, and there are no sanctions against someone who copies another's design. I have mentioned a comment by one artisan who was aware that his designs were being reproduced in large quantities in the villages, but also that he could do nothing to stop this. He took this philosophically, describing the situation as 'helping the poor' (I return to the themes of ownership, designs and authenticity in Chapter 6). In some way, this is related to the fact that there is no requirement that a person knocks on the door before entering a workshop. Artisans feel that they are entitled to enter any workshop, and are entitled to see what others are doing. It was first in the Zisha Factory that other people's work became visually available to all, which resulted from the openness of factory space. Significantly, then, erecting barriers in the factory in the early 1980s to create individual workshops (to form the *yanjiusuo*, or research department) was a way of restricting visual access. Subsequently, it became increasingly possible for artisans to work in the privacy of their home.

Although knowledge is open to the apprentice, it is later on that more specialized techniques are concealed from colleagues. Artisans might stop their work as another enters, so as not to let the other see work in progress. I first became aware of this practice during a conversation with Mr Yi, who recalled how his master would stop working when a colleague he disliked would enter his workshop to prevent the latter from learning about his methods. This was at the time when the communal workspaces had been divided into individual workshops for masters in the 1980s, which gave Yi's master some degree of control over who had access to his work.

A particularly useful tool for this tactic is the vat that artisans use to restore moisture to pots that are being worked on. This large ceramic pot (increasingly replaced by a lidded plastic box) is lined with bricks, part-filled with water and covered with a wooden lid. Artisans usually work on several pots at the same time, and occasionally leave work in progress in the vat while they work on another object. Putting away a pot with the excuse that it is too dry to work on is one way in which work can be hidden from the inquisitive gaze of colleagues, and this was the trick Yi's master would resort to. As Yi described, this did not involve refusing interaction with the visitor, something that would not have been possible in the context of a factory, where interactions were expected between colleagues and secrecy discouraged; on the contrary, there is the potential for the gesture to be interpreted as politeness, interrupting one's work for a chat. No doubt, when a visitor enters a workshop, certain elements are available to see – the artisan at work, the work in progress and the tools being used. However, artisans can act on what is available to another's gaze. I have occasionally seen Yi stop and set aside his work, seemingly to chat with friends, yet not showing much enthusiasm in these interactions. I cannot say what the motivations were for these acts, but it is at least possible that he was following the strategy of his master, by preventing some colleagues from getting to know too much of his work.

Another artisan working in the factory building, Xiong, was known to hide his work from colleagues as soon as they entered his workshop, allegedly to avoid comments that he assumed would be negative. Xiong, however, seemed to be on perfectly friendly terms with many in the factory building. I myself visited his workshop a number of times and had friendly interactions with him, though as predicted by others, he would stop his work when I entered. Whatever the reason for this action – avoiding comments or preventing others from having access to his methods – it indicates that there are reasons for artisans to conceal their work from friends, and that it is a known fact in the community.

Irony was another way in which Yi exerted control over the circulation of information. Using irony enabled him to avoid giving advice whilst maintaining social relations. In exchanges with one particular colleague for instance, his advice would be confined to joking comments on the artisan's products being 'beautiful', which at the same time enabled him to avoid giving any concrete advice or feedback. In fact, I knew that Yi did not think much of the products of his colleague, yet this was one individual with whom he had the most regular exchanges. As I have suggested, there is a fine line between casualness and disinterest in the interactions between artisans, and this is one way in which the efficacy of exchanges in terms of opportunities for learning can be limited.

Such attitudes and possibilities for the curtailment of opportunities for learning have largely been facilitated by changes in the configuration of space in Dingshu. During the first decades of the factory, spaces were open: there were not supposed to be secrets between workers, and the arrangement of space made it difficult indeed for secrets to exist. Everyone's work, including the work of masters, was available for everyone else to see, which created opportunities for others to learn – although there were few incentives for workers to innovate through such learning. Later, in the 1980s, spaces within the factory were divided up for masters to have their own workshop space. The incentive was for masters to develop new products, but a consequence of that move was the reintroduction to some extent of possibilities for secrecy.

It also led to the reintroduction of a form of one-to-one apprenticeship, so that these secrets could be passed on to select individuals. So even if spaces where not private at the time, masters could to some degree control (visual) access to their work. More recently, after the economic reforms, the privatization of spaces of production has greatly facilitated the possibilities for artisans to conceal their work from others, especially for those artisans who have established workshops in their own homes. But as I mentioned, artisans positively associate the changes brought about by the factory with unprecedented possibilities for exchange. It was the negotiation of the two demands of social interaction and of competition that resulted in the particular configuration and distribution of knowledge among artisans.

Borrowing the terminology of Cavalli-Sforza and Feldman (1981) in their discussion of cultural transmission, one can say that there is a tension at play between vertical transmission (from parent to child or within a family) and oblique or horizontal transmission (from an expert to an unrelated novice or between peers respectively). The authors predict that a vertical transmission system results in the most conservative environment, whereas the greater the horizontal opportunities for transmission, the greater the scope for innovation. In this sense, the *yijia yihu* system was to a large extent vertical – households acted on limiting access to resources – and conservative, since they encouraged the upholding of styles specific to the household. The factory, in contrast, introduced a system that was geared towards the oblique and horizontal transmission of knowledge. This did indeed have the effect of fostering innovation: whereas before the cooperative, apprentices usually had one master and based their own career on the master's style, in the factory students had many teachers and increased possibilities for exchanges, and there was thus room for learning a diversity of styles. As a result of that situation, and of the continuous exchange of expertise, present-day artisans have developed more personal styles. This same situation has at the same time brought about the necessity of limiting access to certain methods, since in today's zisha 'art world' it has become important for artisans to be associated with an individual and exclusive style.

As Hutchins (1995) argues, cognition can be distributed in time and space, and individuals can to an extent learn through interaction with their socially constituted environment. The (economic) metaphor of distribution seems suited to describe what is happening in this community of practice. Since, as Bourdieu (1986) points out, the cultural capital acquired through practice can be transformed into economic capital, an economic metaphor need not seem out of place. Artisans might not be fully successful in acting on the possibilities for learning by concealing their work, but still their strategies are efficacious in acting on the distribution of knowledge. Speaking of commodities, rather than practical knowledge, Arjun Appadurai (1986) has described slowing down the flow of things as a manifestation of power within regimes of value. I am not suggesting that commodities can be equalled with skills and techniques; rather, I mention Appadurai here to suggest that one can find alternatives to the metaphor of community in attempting to understand the relationship between social groups and learning within these groups. It is clear to me that there is an unequal distribution of practical knowledge in Dingshu, with individuals attempting to acquire certain skills or techniques, and colleagues not necessarily enabling this acquisition to take place.

For Barth (1969), groups are defined through exclusion rather than through shared identities, a point acknowledged by Wenger (1998), who suggests that in the specific context of communities of practice it is the shared understanding, the shared experience of the members of a group that exclude others from it. In the case of Dingshu, it is clear that the community of former workers at the factory distinguished itself from, for instance, artisans working in the surrounding villages, who did not have the pedigree of the factory and as a result could only produce mere 'commercial' products. Although it was not always the case that former colleagues from the factory remained in close contact, a community was at least expected to exist, it was 'imagined' (Anderson 1983) in terms of the boundaries of the factory compound, even if these no longer contained the community. However, the 'community of sociality' of artisans that was created by the factory is not coextensive with the 'community of practice' defined by the distribution of knowledge within the group, since artisans are acting on this distribution.

The state education system and apprenticeship

During my return visit in 2012, I realized that much had taken place over the previous four years or so in terms of access to the craft. One of the vocational schools in the township had resumed a course in *zisha* pottery that they had interrupted for the past ten years due to lack of demand, and was accepting around twenty students every year (Figure 8). To obtain their degree, students pursue two years of training in the classroom, and a third year with a master outside the school, who reports back on the progress of the student to the school. Is this a return to the state-controlled teaching structures that were created in the Zisha Factory and abandoned in the mid 1990s? Master Wang Yinxian, who perhaps has become the most important *zisha* master in recent years after having been named as an 'inheritor' (*chuanchengren*) of intangible craft heritage at the national level, provides guest lectures on *zisha* appreciation (but does not actively teach techniques), and has provided a drawing of a design of teapot that serves as the model for the one that all second year students reproduce. In a photograph posted on the website of the

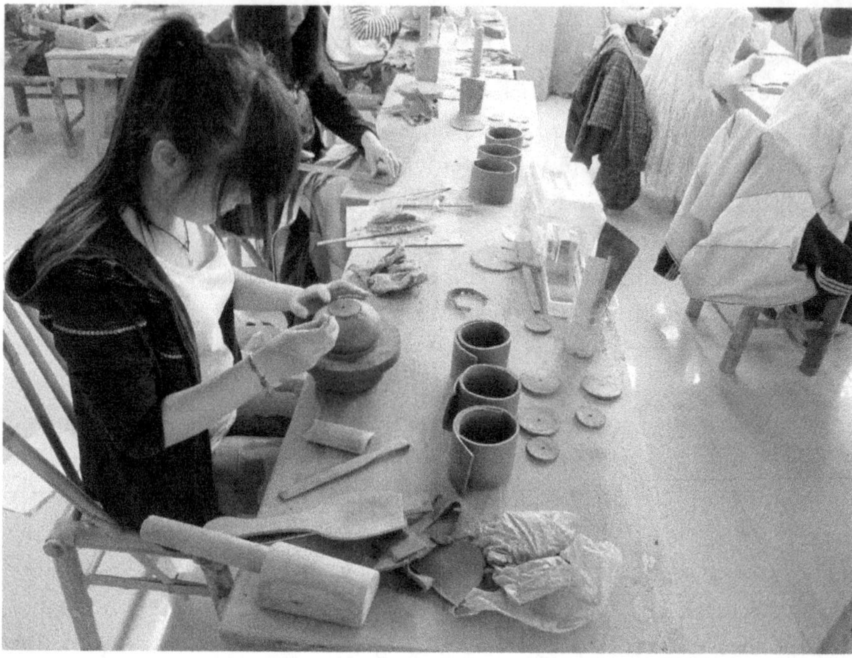

Figure 8 *Learning to make* zisha *pots in the classrooms of the Dingshu vocational school.*

school, the master is posing with a teapot in her hand, pointing to significant features of the pot, with three students looking on intently; in the background, a teapot design is drawn on the blackboard. This photograph is reminiscent of black and white photographs from the 1950s to 1970s, with masters and students adopting the same poses as those in the contemporary photograph.

There are, however, important differences between an apprenticeship in the former factory and learning in the vocational school. Students have to spend their third year in gaining 'work experience'. This in practice means that they are apprenticed to a master outside the school. A survey of twelve students in their third year pointed to the fact that all the students had contacted their teacher in the same way as students who learnt away from the vocational school: through the network of contacts of their parents. In several cases, their teacher was a member of their family; in other instances, students paid private tutors to learn the craft. Private tuition was referred to by one of my informants as a new 'industry' (*hanye*) that had appeared, he suggested, around 2008. In one such establishment, some twenty students were learning the craft, and most told me that they would stop after their first year and rely on the networks of their parents to continue learning with another master. The increase in demand for learning in Dingshu, an informant suggested, had made teachers more selective, and a student who had already acquired basic knowledge of the craft had better chances of getting accepted by a teacher. Another new strategy employed by teachers was to accept up to about four students simultaneously, and then agree to continue teaching only the most talented one beyond the first year.

The significance of the vocational school in transmitting craft knowledge is still to be determined. Many of my informants dismissed it as unimportant, pointing to the proliferation of private structures of learning and to the fact that learning in a classroom cannot replace a traditional apprenticeship. At least symbolically, however, the revival of state vocational training in *zisha* is a sign of the new relationship the state has established, or re-established, with the craft.

Conclusion

The two 'modernities' of collectivization and the subsequent turn to capitalism had a number of consequences for the distribution of work and knowledge in Dingshu. I have argued that collectivization had as an effect the redistribution of the practical knowledge associated with the craft of making *zisha* ware. Previously secret knowledge, passed on from parent to child, was taught to large groups of students. Students were recruited irrespective of class, background and gender, and many individuals who previously did not have access to the craft joined the factory as apprentices.

Narratives about the creation of the Zisha Workshop stressed the break in the closure of knowledge (which characterized the *yijia yihu* system) that came about when masters and a generation of new students were regrouped under the same roof. That communal era has become an idealized moment when social ties were created seemingly out of nothing. Whereas in the past a master taught one, sometimes two, students at a time, after 1954 every apprentice had twenty classmates, who would give each other support, and solidarity between them would remain after the end of closure of the cooperative factory. It is through the present exploitation of these ties that artisans can perpetuate some of the services and advantages that the now-privatized factory no longer provides, in particular teaching novices and access to knowledge.

Despite what one might have expected, the model of the factory was evaluated in generally positive terms. Not that there was a desire to return to this state of affairs, but it was recognized that the present-day dynamism of the craft was the result of the extraordinary social potential that the factory had created: the breaking down of barriers to learning, and the creation of a new kind of community. This notion of community continued to be fostered in the context of regular exchanges between artisans, based on the movements of *lai baixiang*. But I have also shown that there are a number of ways in which artisans negate the existence of any 'community' values. This is not so much by keeping secrets but by working on the very basis upon which social ties rest, movements related to the notion of *lai baixiang*, as well as hiding their work in progress or refusing to see someone else's work. I have rooted the possibilities for the movement of expertise and services in the new context created by the factory, and especially the 'revolution' brought about in the way the craft was taught. I often wondered about the economic incentives of learning: apprentices did not produce anything that could be sold, and did not render services to the master, besides maybe preparing tea and cleaning the floor in the morning. I would suggest that the key to that puzzle is social rather than economic. Teaching creates and maintains social bonds, between apprentice and master obviously, but most importantly between masters and the parents of apprentices. The apprenticeship system in Dingshu is caught in the system of mutual exchange of favours that regulates much of Chinese social life. Crucially, this movement of *lai baixiang* is connected to the act of seeing, and although artisans do not usually try to prevent someone they dislike from visiting their workshop, making the process of work unavailable is one way of preventing elements of techniques from reaching those whom they believe are not worthy of knowing them. Recalling Dilley (2004) and his comments on Appadurai (1986), one could say that in this way artisans act on the circulation of knowledge by preventing visual access to certain methods of production in the 'regime of value' that regulates the exchange of (valuable) ideas, techniques and designs.

At this point I wish to return to the case I started with. Wan told me he was not too happy about his two apprentices, who both came from potting families. He accepted them out of feelings of obligation, but did not enjoy teaching students who had started learning with others, that is, who were not entirely his 'products'. He was imagining what he would be doing in ten years time, teaching a class of twenty novices. As far as I know, and from my interpretation of the attitudes of people, Wan is the only person to have such a plan in mind. Teaching many students at once has the advantage of creating solidarity among students. If a lesson has been misunderstood, students can help each other. Wan himself recalled his years of training in a class of twenty students and the feelings of solidarity between classmates.[4] In fact, in his description of his years as an apprentice in the old factory, students were left in large part to their own devices: the teacher would explain the process once,

4 Lave and Wenger suggest that there is 'anecdotal evidence' that 'where the circulation of knowledge among peers and near-peers is possible, it spreads exceedingly rapidly and effectively' (Lave and Wenger 1991:93). Wan, and other artisans, seem to concur on this point.

then leave students to repeat the movements, and they would occasionally go through the ranks to control what was being done. Wan was further thinking of the strong relationships he had developed with his former classmates, and was referring explicitly to the situation of the Zisha Factory and his aims at reproducing those same conditions.

A second important aspect of Wan's project was the teaching of students who were absolute novices. This might not sound a productive idea, since Wan could not know whether among these students there would be one talented enough to be his successor (and as we have seen, talent is not developed through hard work). Would it not be a better idea to take on students whom Wan would know to be talented? In a way, Wan's idea implied an interruption of the otherwise apparent continuity with the factory years: his ideal students were those who would only have had him as a teacher, and in this way Wan would be creating his own 'community' of students.

This suggests to me an intriguing parallel between the making of a pot and the 'making' of an apprentice. Artisans have always rejected a division of labour, and insist that an individual artisan has to complete every step of the process of making a pot, even the most basic ones, so a master has to take care of all the phases of the making of an apprentice, from the very start. Wan saw himself as the keeper of a tradition and deplored other artisans who did not properly follow the principles of the craft. In teaching students in his own way, Wan was planning to cut out all influence from those he perceived to have less rigorous teaching methods. Cutting the network (Strathern 1996), denying access to others, thus becomes a way of preserving the integrity of the craft, whilst at the same time improving its value. One might follow this parallel between apprentices and pots further, by asking whether the making of *zisha* pots is also about cutting networks, and if so, what is being left out?

CHAPTER 5

Engaging with the craft

Artisans in Dingshu take pride in the belief that their craft is unique in the process, techniques and tools it uses. In this chapter, I pay closer attention to the techniques of *zisha* pottery. The aim is not simply to give a better appreciation of the craft and its techniques, but also, in the second instance, to address one aspect of the 'exclusivity' of the craft, its unique technical process, that is actively and creatively maintained by artisans in their daily work. In talking about the techniques of *zisha*, I draw from observations of artisans at work, their explanations about elements of techniques, as well as from my own personal experience of learning the craft with an artisan working in the buildings of the former Zisha Factory. I introduced the topic of apprenticeship in the previous chapter, and I continue with this theme here, in relationship to my own 'apprenticeship', during which time many elements of technique were revealed to me, elements that cannot be understood easily with the simple observation of an artisan at work. Techniques, tools, skills, talent, body posture, aesthetic principles – all contribute to the making of a *zisha* pot. Becoming an expert involves apprehending all these elements at once in one's engagement with materials, tools and the physical space of the workshop. Apprenticeship is a process in which the various elements of technique are at first learnt independently, and later comprehensively mastered to create a craft object from start to finish. In the previous chapter, I uncritically talked of the 'transmission of knowledge'; this phrase is however fraught with debate in the anthropological literature on apprenticeship and enskilment. As many have argued, tacit, practical knowledge cannot be conceived of as a stock of knowledge that can be passed on to the next generation, as if skills were independent from the master or learner and learnt as formulae (Bloch 1991;

Ingold 2000; Marchand 2003; Palsson 1994; Polanyi 1962). I adopt the notions of engagement (Ingold 2000), and dwelling (Polanyi 1962) in activities, which point to a greater agentive role on the part of the learner. Beyond these approaches however, I want to think about the relationship between processes of enskilment and the development of a sense of artisan identity.

In Dingshu, the first months (and up to a year) of an apprenticeship are involved in mastering the making of the building blocks of the craft: the flat strips and discs of clay shaped and flattened with a mallet. After these basics are learnt, the apprentice progressively learns, one after the other, the subsequent stages of making a pot: forming the body, adding the neck and rim, adding the spout and handle, smoothing the walls of the pot. I have written this chapter in a similar linear way, following the stages of the making of a pot, which are also the stages of an apprenticeship. I identify two fundamental principles of the craft – sequence and precision in movements – which once mastered together transform the apprentice into a proficient artisan.

The chapter is also an occasion to reflect on the impact of technical changes – in particular the use of what are referred to in Dingshu as 'modern' (*xiandai*) tools – that have changed the relationship the artisan has with clay. Modern tools simplify the work of craft of making a pot, but in so doing they also standardize production and create a distance between the artisan's body and the clay. This discussion leads me to pursue a narrative thread that I began earlier in the book, on visual access to methods of production. Here, I look at the qualitative dimension of visual access, and suggest that vision should itself be understood as embodied. I start by characterizing the craft in terms of the relationship between the object in formation and the artisan's body in terms of a 'negative' of sorts.

The negative pot

Artisans in Dingshu like to stress the uniqueness of the techniques of their craft. This claim might not be entirely true: *zisha* is not the only form of ceramics that uses leather-hard clay to slab-build pots. Still, it has developed a very precise and specific panoply of methods and tools that are most probably unique. To understand these rather unusual techniques, it is useful to contrast them with wheel-throwing methods. Often artisans themselves resort to such comparisons and contrasts when they explain their craft since they realize that wheel-throwing techniques are the most well-known methods of ceramic production, both in China (where porcelain is mostly worked on a potter's wheel) and in the West.

Zisha makers mostly work with leather-hard clay – clay that is relatively dry, wet enough to be flexible, yet not dry enough to break when bending – in contrast to wheel-thrown potters, who work with relatively wet clay, elastic enough to be directly shaped with the hands. One of the consequences of this difference is that making *zisha* pots is not a particularly messy activity: the workshops of *zisha* artisans are almost invariably immaculate. Using leather-hard clay means that the clay does not splutter, as it does when a potter works on the wheel. Even the artisan's hands are clean throughout most of the process of making a pot. Artisans have a near obsession with cleaning – tools, hands, table: all need to be wiped as often as possible, an action further reproduced throughout the process of making the pot, as the surface must be repeatedly smoothed with different tools. This is largely the requirement of a craft that both needs to appear virtually flawless and does not use glaze to cover any irregularities in the clay. A speck of dry clay on the otherwise perfectly smoothed surface of a pot, for instance brought about by a tool that has not been properly cleaned, can immediately devalue the object. An apprentice is constantly reminded by their master to *ca ganjing* ('wipe clean') tools and surfaces during and after every procedure.

The basic building blocks of *zisha*, as I have already mentioned, are clay discs (*nipian*) and clay strips (*nitiao*), about 2 or 3 millimetres thick, which are shaped by beating with a wooden mallet with a flat, smooth surface. The act of beating the clay into shape has the added advantage of compressing the clay, making the final object stronger. These two elements, the clay discs and strips, make up, through different processes, all elements except the spout, handle and knob.

The profile of a *zisha* pot is a continuous line. However, in contrast to wheel-thrown ware, this is not achieved from a single lump of clay; the profile is rather made up of many different elements added together. The part that forms the foot ring, or the neck of the pot, is actually an element that is added to the piece of clay that forms the body. The appearance of a continuous line is the result of patching up with slip (a paste of wet clay) added to joints and carefully smoothed and shaped. For instance, the neck of the pot is made by stacking two discs of clay on top of the already formed body, which itself is closed up by a third disc. Slip is added at the point where two discs meet, and shaped with a tool to form the final profile of the pot. Towards the end of the making process, the centres of all three discs are cut out, leaving only a rim and creating the mouth of the pot.

This 'additive' aspect of the craft forces it to be a rather precise affair, for all elements must be cut to precisely the right dimension before they are added to the pot in formation. Artisans use a cross-section plan of the pot's design, or a schema representing all the measurements of the pot, and refer to them to precisely measure the planned dimensions of the pot with a tool, the *juche* (Figure 9), a kind of compass-like cutter made of bamboo with a metal blade, which is then used to cut out the desired shape out of the clay disc. With

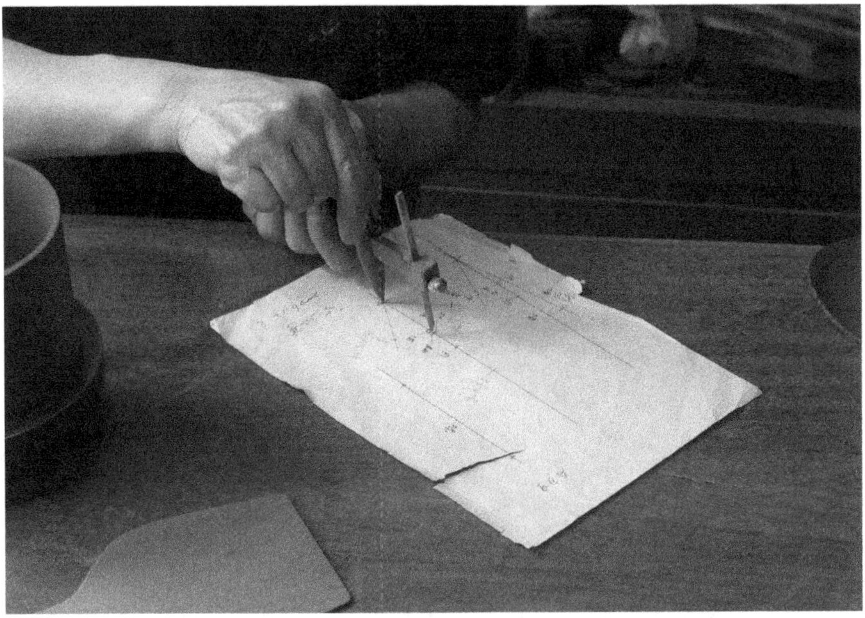

Figure 9 Taking exact measures before cutting the clay (this and Figures 11-15 show artisan Zhao Jianghua at work).

zisha, hands seldom shape the clay directly, and artisans have, across the ages, developed a range of tools, each for a very specific function.

To give an idea of the vast array of available tools, several of them are used to simply give a finish to the holes inside the pot that serve to filter the tea; these holes will not even be seen in the finished product, yet the perfectionism of the craft (at least of the 'elite' artisans producing highly valued pots) dictates that they be regular and given a proper finish.[1] The same amount of attention and rigour employed in the making of the pot is used in the making of tools. Any good artisan must be able to make their own tools, or at least modify to requirement any purchased ones. Tools are made of bamboo, different kinds of wood, and buffalo horn, the latter particularly useful to smooth the surfaces during the final stages to give the clay the sheen that is characteristic of *zisha* pots.

One might identify the *bizi* as one of the most characteristic of the tools used (Figure 10). Made of a section of bamboo, it is usually employed to work the 'roundness' of the teapot: the artisan is exploiting the natural round line of

1 This also gives an indication of the high demands of purchasers, as even such details influence the final value of the object. I have seen buyers looking at a pot through a magnifying glass to pick out any irregularity. Of course, this is true only of the more highly crafted pots.

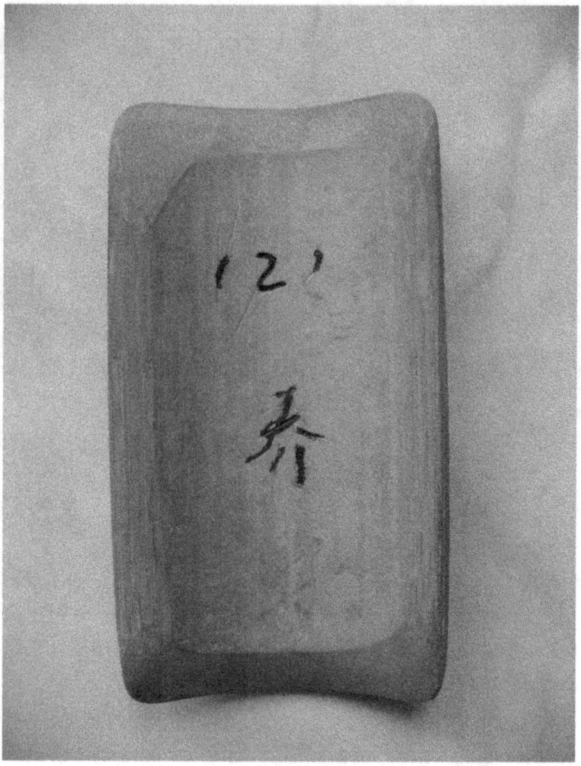

Figure 10 Bizi *tool made of bamboo. The natural curve of the bamboo has been modified at both ends, and each will shape different parts of a specific design of pot. The Chinese character, Qiao, is the Chinese name of the author, for whom this tool was made.*

the bamboo section, but will also slightly modify this line by filing it in order to give it the desired line for a specific design of pot. This tool is lightly pressed on the surface of the pot, and through repeated turning it gives an imprint of its line. The shape of the tool is complementary to that of the various parts of the pot. The two ends of the tool in fact operate on different parts of the pot: one end has a slight angle in the line, and shapes what is known as the 'shoulder' of the pot, and the other end shapes the pot's 'belly'. Interestingly, these tools are, ideally at least, made to specification for each type of pot to be made because the shape of the tool must correspond to the lines that are to be given to the pot. Few artisans actually follow the principle of making new tools for different pots, but those who do start by drawing a precise cross-section of the design, and from this they go on to make the *bizi* tool. The latter will, however, have to be modified during making to correspond exactly to the line the artisan has in mind.

The lines of the pot are, so to speak, already contained in the lines of the tools. One could think of the tools as 'negatives', in a photographic sense, of the pot, as they contain the concave line that in negative will form the convex line of the pot. This is all the more true of one of the most recent additions to the panoply of tools at the artisan's disposal, the plaster mould, which contains the whole pot as a negative image. In fact, I feel that the metaphor of the negative can be extended to the process as a whole. I have in mind an image of the artisan working over a pot that turns on itself and remains centred whilst all movements of the body are directed towards that central point: it is the whole body that becomes this negative. Of course this might also describe other methods of ceramic production, such as coiling and throwing techniques. But *zisha* might illustrate the point a little more forcefully, by virtue of the types of tools it employs. Another contrast to throwing techniques, in which the artisan's hands move with the clay in slow movements, is that when making *zisha* the clay is beaten into shape with quite vigorous movements – in stark contrast to the small size and fragility of the clay object taking shape. When beating the clay, the artisan must each time strike it at a precise point, and what's more, it must contribute to the clay staying in the centre of rather than flying off the table.

If the body is to act as a negative of the pot in formation, its movements need to be precisely controlled – it needs to be correctly positioned, one might even say 'constrained'. When I started learning the craft, the first thing I was taught was how to sit at the table properly. My teacher sketched for me an overhead view of the correct position: the arms have to form a triangle, with the pot at the apex, the elbows forming the two other corners. The arms must rest on the table, which means the chair and table must be of the correct height. It was important, I was told, to be comfortable at the working table, yet many of the positions I was told to adopt did not in any way feel comfortable. The first technique an apprentice learns is the making of clay discs, *nipian* (also known as *qiezipian*, or 'aubergine slices'). These discs are shaped from a cube of clay beaten with a mallet with a flat surface and cut out from the resulting flat clay; they are used to form the base and elements of the neck and lid. The ideal thickness is 3 millimetres, though an apprentice will be told to make them slightly thicker to get them regular more easily. Some smaller pots, valued for their lightness, require that the clay be only 2 millimetres thick.

The exercise of beating the clay is a feat of movement coordination: a wooden mallet is held in the right hand, and strikes the clay in an up and down movement. Whilst the right hand is in the air, the left hand lifts the forming clay disc from the table by a few millimetres, turns it anti-clockwise a few degrees, and replaces it on the table for the next strike. The coordination of the circular movement of the left hand and vertical movement of the right is not easy to achieve for a novice such as myself: the right hand tends to want to go in circles too, or the opposite. The next technique is the making of clay sheets, the strips (*nitiao*) that will form, when curled into a cylinder, the body of the pot. Here the same mallet is used, but held in both hands, striking the clay away from the centre in order to flatten it. The technique is a little less demanding in terms of coordination, but no less difficult to master.

These two techniques can take a long time to learn properly. In the Zisha Factory, in some cases as much as a whole year was spent practising these movements, while novices today spend at least several months practising them. Considering that the full length of a traditional apprenticeship is three years, this is a relatively long period of time to spend on one part of the technical process. It is not that the movement itself is so difficult to master, but it takes a long time to learn to get the discs and sheets absolutely flat and of the right thickness. This is often checked by the master by piling the sheets of clay one on top of the other and evaluating the flatness of the resulting stack. After evaluation by the teacher, the student is told to cut the clay into pieces, throw the clay into a bucket of water in preparation for the recycling of the clay the next day, which will serve for practising the same exercise. Needless to say, this can be a tedious and at times frustrating practice. The exercise teaches coordination, skill and patience. But it had an extra function, at least in the old factory, that I hinted at in the previous chapter: in the mastering of these particularly difficult skills was decided much of the fate of apprentices, and those who were not able to properly make these clay elements were made to stop and transferred to other activities.

In order to teach the correct movements, a master starts by demonstrating the technique, with more or less verbal explanation (depending on the teacher). It is important, I was told, to be comfortable at the working table. However, I was asked to adopt some uncomfortable positions. In particular, tools were to be held at awkward angles. When I attempted to reproduce the movements as they had been demonstrated, my teacher would often come to stand behind me, grab my hands and squeeze them into place, to get me to execute the correct movement. I was interested to see a similar teaching method used by parents to teach their child to write Chinese characters; the parent would grab the child's hand, and move it to write the character. The difference of course was that in this case, only one hand was used. Other apprentices confirmed that this was a common method of teaching. However, it does not last throughout the apprenticeship period, and during my three months of learning I noticed that there were fewer occasions on which my teacher would directly hold my hands to get them into place. Former apprentices confirmed that after about a year, teachers direct the novice's movements almost exclusively by demonstration and verbal explanation rather than physically guiding them. This shift in teaching method indicates that this stage of 'positioning the body' has been completed.

In verbal instruction, repeated phrases not only guide the apprentice but also give a sense of rhythm to the process. In a procedure whereby I was to shape the body of a pot, my teacher would repeat the phrase *nong gao, yaxiaqu* ('make tall, push down'), which gave an indication of the alternating movements that would properly shape the pot: pushing up the walls of the pot with the *bizi* tool, then pressing down at the centre of the top of the pot in order to make that part of the pot flat. The place of verbal instruction might be surprising, given reports from other anthropologists of craft who have pointed out the relative absence of verbalized instruction (e.g. Lave 1977; Marchand 2001; Singleton 1989). In Dingshu, this might at least be partly a legacy of the factory years, when verbal instruction became important in the context of classroom teaching. It is also cultural, and I have noted similar regular recourse to verbal instruction (again, used both as instruction and rhythm setting) by a master ceramicist in Taiwan who had received instruction from his father in a classic one-to-one apprenticeship.

Feeling and seeing

Three years was deemed the appropriate time period for an apprenticeship, even if many apprentices spent a shorter time learning the craft. This period of apprenticeship is common throughout China. A couple of artisans who had just completed their apprenticeship period explained that these three years were necessary to learn the 'nature of the clay' (*nixing*). This involved mastering a number of technical aspects, including knowing what action needs to be performed according to the state of dryness of the clay, and mastering the individual techniques of the craft. But more generally, learning the nature of the clay involves developing a 'feeling' (*ganjue*) for the characteristics of the clay. *Ganjue* is the generic term used in Dingshu to explain the development of a skill through practice. The following example can illustrate the notion of *ganjue*. One day, Mr Gua walked into the workshop where I was learning *zisha*. I was attempting to fit a base on the pot I was making, but found that the disc I had beaten had not been cut to the correct dimensions. To help me out, Gua cut out a piece of clay and made a new disc for me. Initially I was impressed with the speed at which he made this, and then by the fact that he had reached the correct thickness of 3 millimetres within seconds without stopping to look. Perhaps most strikingly, the piece of clay he had cut out seemingly without thinking was just a fraction bigger than the size of the disc I needed. My teacher joked, 'this is why he has the title of "advanced craftsman"', and went on to explain that this was *ganjue*, the ability to get it right, knowing exactly how much clay to use and getting both the right size of disc and beating the clay to exactly the right thickness. Gua emphasized the point by rubbing his thumb and forefinger together, also implying that he could feel the thickness of the clay in his fingers (through the left hand that turns the piece of clay around as it is beaten). Arguably, *ganjue* in this case implies not only tactile skills but also visual skills, specifically the ability to judge how much clay to cut out for the desired diameter of disc. I have emphasized the role that vision plays in the learning of the craft, but of course the development of a sense of touch in craft is crucial. However, I discuss the sense of touch relatively little here because in contrast to vision, touch was rarely referred to in discourse. The sense of touch is to be developed through experience and dedication by the apprentice and is relatively individual, whereas 'seeing' properly is stressed in the instructions of the teacher.

Vision plays an important part in learning the craft, and in more complex ways than in mere observation and imitation. I have mentioned that the learning process evolves from more direct interaction to control at a distance through words. These two methods, direct control of the hands and control through the voice, are modes of teaching through 'scaffolding' (Bruner *et al.* 1976), meaning that the tasks to be learnt by the apprentice are made more accessible through the intervention of the expert who directs the student's efforts. Another substantial part of learning does not take place by recourse to scaffolding; rather, techniques are made visually available to observation, making it possible for learners to pick up techniques without explicit instruction nor direct intervention (see Marchand 2007). The physical proximity of apprentice and master in the workshop was often pointed out by my informants as essential to the learning process. Usually the work table of the apprentice is placed in close proximity to that of the master, which enables mutual visual access to the work of master and apprentice. The teacher will quickly spot a mistake being made by regularly glancing up from their own work, whilst the apprentice can easily raise their gaze to observe their teacher at work and strive to better understand a certain technique.

There is a certain quality that is implied in the act of looking. An example will illustrate this point. When I asked an artisan, whom I had filmed at work, whether I could publish the film on the internet, I was given the green light. I tried to push the issue further by asking why he was not scared of someone learning secrets from his work (the film after all showed in detail all the steps of the process of making a pot). His answer was that this is impossible simply by looking at a film: it is necessary to have learnt from a master. What more does learning from a master involve than simply looking at what is being done? He gave two answers to this. Firstly, it is important to have some prolonged visual contact, hence the importance for the apprentice to work beside the master. The second reason is that there is feedback from the master, who is also able to look up from what they are doing and point out a wrong action.

One might recall here the act of concealing one's work as a way of cutting flows of knowledge, as discussed in the previous chapter. This act should be understood not simply as the hiding of information – designs, tricks and so on – but, importantly, refusing the qualified visual access that is necessary to learning the tacit knowledge of a craft. Interestingly, the way village artisans usually explained to me their 'apprenticeship' contrasted with what I heard from those who learnt in town, either in the Zisha Factory or in private workshops. They would tell me they had learnt the craft 'by themselves' (*zixue*), 'just by looking' (*wo bian kan, bian xue*). In this expression, they were stressing what they perceived as the relative ease with which they had learnt the craft. But more crucially, they were referring to the context within which they had learnt. In some of these villages, virtually all the families are engaged in the making of *zisha*, so that someone living there picks up the techniques of the craft simply through daily interaction with friends and family. Learning to see is an integral part of the process of apprenticeship, and continues throughout an artisan's career. I have mentioned that towards the end of my fieldwork my teacher told me that I was starting to see the worth of objects: I could recognize, at least to some extent, which were good pots. Being able to see is what differentiates amateurs and novices on the one hand, and connoisseurs and experts on the other. In this way, vision, in contrast to for instance touch, gives way to judgements of value.

Michael Polanyi (1962:55) argued that connoisseurship is not made up by a set of rules, but rather only transmissible through example, and therefore a skilled practice. Cristina Grasseni defines her concept of 'skilled visions' as 'specific capacities of perception and action [that] develop thanks to participation in a perceptually and socially rich field of relationships' (Grasseni 2004:49). In Dingshu, the specific visual capacity to see in a certain way starts to be developed through apprenticeship, and, as described in the previous chapter, is developed through participation in the community of craft practitioners.

I would like to further qualify the vision that artisans develop with reference to the structure of the Chinese language. *Kan* is translated as 'to look' or 'to gaze'. The verb 'to see' is usually rendered using the verb 'to look', and adding either the word *jian* ('to see') or *dao* ('to arrive at'), using the particle *de* between the two words. This particle in Chinese linguistics has no equivalent in the English language. *De* expresses either capacity or result: the verb that follows the particle is the result of the first. In our case, one might translate the expression *kan de jian* as the 'capacity to see by the action of looking'. This sounds redundant in English, but makes apparent the chain of cause and effect of the act of looking.

The word *kan* can mean both 'to look' and 'to read'. There results something of an ambiguity then between looking at or 'reading' a work of art – something that some anthropologists writing on art, in particular Gell (1998), might find inadmissible. This of course goes together with the ambiguity of the Chinese character as word or pictorial representation. *Kan de dong*, which translates as 'understanding', but more precisely 'understanding by looking', can refer to reading (either generally being literate or understanding a piece of text), but can also refer to being able to 'understand' a work of art. I often heard the word *dong* ('to understand') with reference to the ability to understand art, but also, in conjunction with *kan*, implicitly connected with the act of looking.

To give an example, one artisan told me on more than one occasion that he did not understand Western art (*wo bu dong*), for instance referring to the artistic ceramics of Western potters he knew from a book, or to the paintings and ceramics of Picasso. Conversely, in other instances, he would tell me that, because of his experience, he could understand a *zisha* pot – know its aesthetic value – just by looking at it: *yi kan yi dong* (which might be rendered literally as 'once see, once understand'). In this last expression, there is a collapsing of the separation between the act of seeing and that of understanding, as if, for an expert, the very act of looking implied full comprehension.

Sequence

Once the basic movements are mastered in the first year of apprenticeship, teaching develops along what I would identify as two complementary lines, two aspects of the craft that will continue to be perfected throughout the apprentice's career: 'sequence' (*xu*) and *gongfu*, or 'precision in movements'. I start with sequence.

Lave (1977, 1990), speaking of traditional tailors in Liberia, suggests that during an apprenticeship, the sequence of learning, its 'curriculum', is not dictated by purely pedagogic considerations but mainly by economic ones. Apprentices learn to sew before they learn to cut, because a mistake done when cutting can cost a piece of cloth, whereas sewing can easily be undone. The sequence of learning thus does not correspond with the sequence of making a garment, whereby the proficient tailor first cuts the cloth, then sews the pieces together, because economic considerations take precedence over pedagogic ones. In Dingshu, however, apprentices can afford to learn each stage in succession since the only way of wasting the material is to fire it. Clay has this particular quality – it can be worked on at leisure, and can easily be returned to its original state by adding water and beating it to remove all the air bubbles. Throughout the process of learning, therefore, apprentices work with the same material that they shape over and over, wasting only a fraction, clay being in any case a cheap commodity in Dingshu. Certain stages of making are in fact far easier than the initial stage of beating discs and clay. In my short apprenticeship where I was taken through all the stages of the craft of making *zisha* ware in a few months, the making of the spout and handle seemed relatively easy compared to the techniques of beating and patting to form the body of the pot (*da shentong*). Why then is it important for the sequences of an apprenticeship to correspond to the sequence of making a pot? The intended effect is to inculcate the sequence that is so important in the craft – and relatively difficult to remember.

Making an entirely handmade *zisha* pot is a rather complex process, one that involves mastering a range of different techniques (see Figures 11–15 for some key moments in the making process). I have mentioned that even when artisans worked in factories, the production of teapots relied minimally on a division of labour. Even in the Zisha Factory, artisans were expected to carry out the process of making a teapot from start to finish, and it was even up to them to make their own tools. In contrast, other Chinese ceramics traditions, such as that of Jingdezhen, the 'Capital of Porcelain', or that of Yingge in Taiwan, have long relied on a division of labour, notably between the stages of forming the clay object, glazing, painting and firing. In Taiwan's main centre of ceramics production, Yingge, where I also conducted fieldwork, a potter

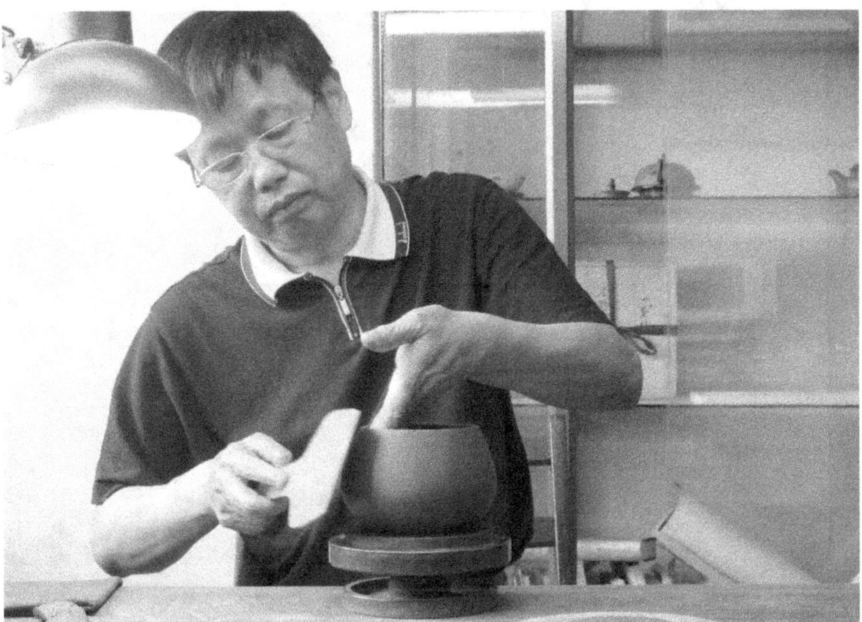

*Figure 11 Patting the body of the pot (*da shentong*).*

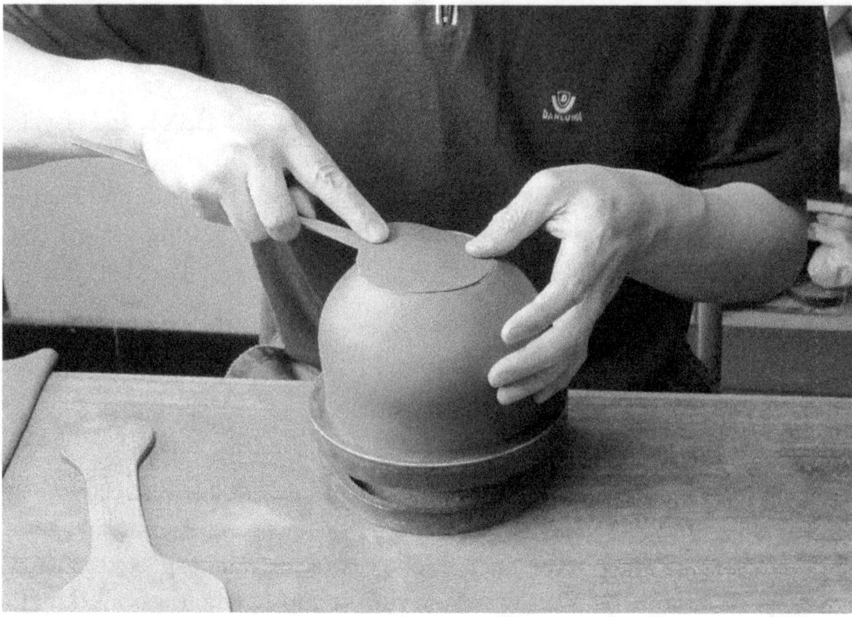

Figure 12 *After the upper half of the pot has been shaped by patting, the form is closed up with a clay disc. The whole object will then be turned over to shape the lower half of the pot in the same way.*

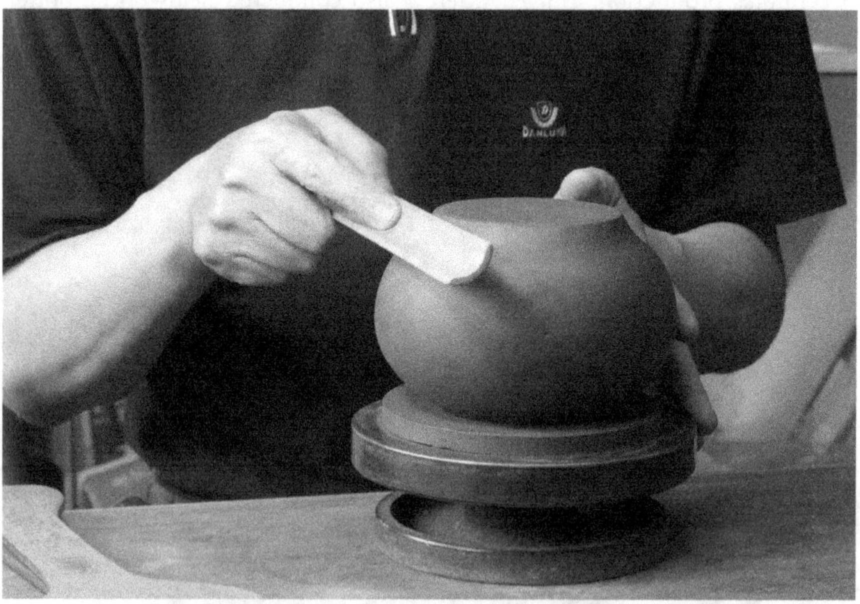

Figure 13 *Shaping the neck of the pot. This technique using a bamboo tool is actually seldom used, a more common way of shaping the neck is to add a ring of clay on top of the body, joining the two elements with slip.*

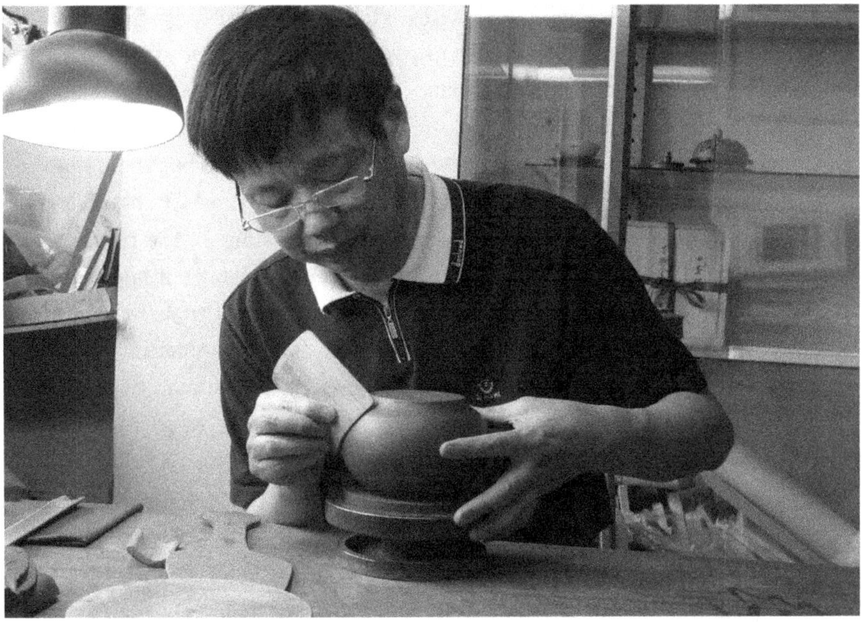

Figure 14 Using the bizi *tool to shape the pot.*

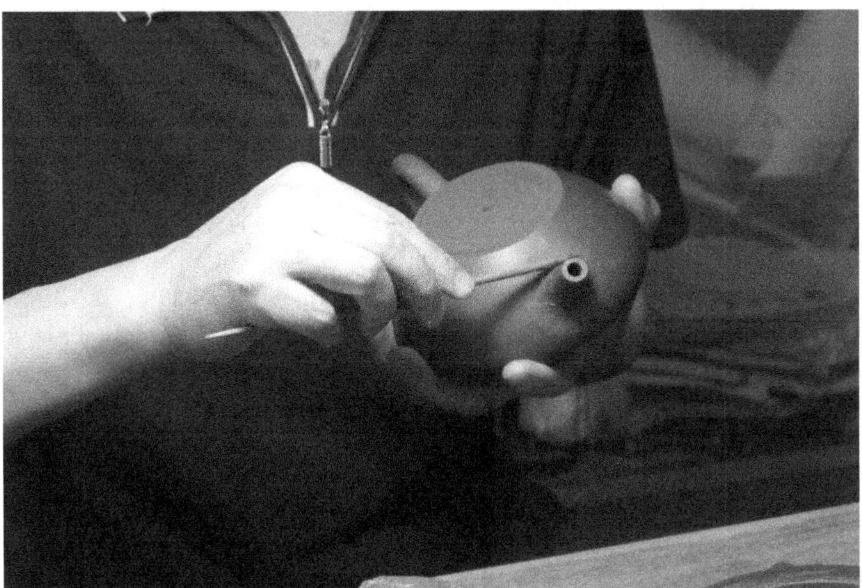

Figure 15 Joining the spout to the body of the pot using slip, and creating a smooth line with a bamboo tool.

needed an assistant (ideally his wife) to 'kick' (turn) the foot-turned wheel for him; they also ideally had sons to finish the pots at later stages in order to produce efficiently. The particularity of *zisha* in this respect, the absence of division of labour in the making of clay objects, was explained to me by pointing out that the quality of a master's pots results also from the quality of the basic stages: making the clay sheets and discs. Only those prepared by the master are suitable for a master's pot, because every stage of the process will be apparent in the final pot and one cannot correct a mistake at later stages.

This explanation half convinced me. I can understand that this might be the case for ceramics of a high standard, but in a factory efficiency is usually a crucial criterion. In the Zisha Factory, most workers were paid by the piece, and it might have thus been important to distinguish between the work of individual workers. But another reason hinted at by my informants is that the artisan needs to be aware of the varying stages of dryness of the clay at any time to know how to work it, or whether to restore its humidity for instance by spraying it with water. Understanding and memorizing the optimal sequence of making a pot, in relation to the stages of dryness of the clay, is crucial for the apprentice to master. Making a pot is at times a race against time. At the beginning of the process, the clay is as hard as leather, and becomes dryer as time passes. This is not an absolute problem, since there are tricks to restore moisture to the pot, either by leaving it in a container with water in the bottom (the pots rest on bricks above the water line), by spraying water directly on the pot or applying water with a paintbrush. But artisans need to exploit the different stages of dryness of a pot to accomplish different operations; they must learn to engage with the material as its state is transforming, and respond accordingly. As Ingold (2007) points out, materials are not inert but 'alive' and always in a state of flux; the artisan needs to engage with this state in order to respond to it appropriately.

To give an example, at the stage of beating the body of the pot into a spherical shape (*da shentong*), it is opportune to finish the process within fifteen minutes lest the clay becomes too dry, for attempts to restore moisture result in cracking. The drying of the clay is also often exploited in certain technical processes. When perfecting the line of the body, using the *bizi* tool, for instance, at first the operation leaves strong imprints in the clay modifying the shape significantly, because the clay is relatively wet. As the clay starts losing moisture, the operation works in a more subtle and precise way. At other moments in the crafting process, the different stages of dryness of different elements of the pot must be taken into account. This means, for instance, that one must work more or less at the same time on the base of a pot and its 'neck', lest cracks appear due to uneven drying. An elegant trick, that exploits the stages of drying, is used when cutting the three discs that make up the neck, the rim of the neck and the rim of the lid. All three are cut with the same tool set at the same length, and thus cutting three discs of exactly the same diameter. However, the three pieces are cut at different stages of dryness. Because the clay shrinks as it dries, the pieces will be of different sizes since the drier piece will not shrink as much as the wetter one, even though they initially all had the same diameter. This trick is used to make the rim of the lid slightly larger than the rim of the mouth of the pot, which is considered more aesthetically pleasing.

Artisan Ge Taozhong told me that his master, Gu Jingzhou, placed great importance on getting his students to memorize the sequence of steps in order to perform all the stages of making a pot with perfect fluidity – almost without thinking. I can illustrate that point with something I noticed when editing a film I had made of Ge making a teapot. I edited other films by more or less experienced artisans, but in the case of Ge, I found it rather difficult to find the right moment at which to cut a scene. It seemed that the very moment he stopped performing one action, his hands had already started to fetch another tool, or his eyes had set on another part of the pot. I was reminded of a comment the Hollywood film editor Walter Murch (1995) makes. For Murch, the edit point in a film is similar to a blink. A blink not only performs essential maintenance of the surface of the eyes, but is also associated, according to Murch, with the end of a thought process. Editing a film is like blinking, ending a thought. Murch describes how he noticed that certain actors would blink at the very moment he would want to cut the film, proving to him that it was indeed the correct moment for an edit. In the film I took of Ge, there are only a few moments where Ge's thought process seems to stop, where he blinks to end one thought and permits another to start. His actions follow one another in a fluid and seemingly uninterrupted process.

In another case, I filmed a younger artisan who had twelve years of experience as a potter. He had learnt from his father, who was present at the time of filming. If one compares this with what I filmed of Ge, the process of making a pot was more clearly a series of individual actions rather than a sequence. In fact, the father was at times giving directions, which mostly had to do with the proper sequence: he would remind his son of an action that he had forgotten, or told him how to proceed. The son had enough experience to finish the work by himself, but doing things in order might better guarantee that a pot came out without any imperfections, for instance that an element is not left to dry too long before it is worked on. I was usually told that apprenticeship should last three years, but this last example shows how learning continues after this initial stage. If after twelve years this artisan was still receiving instructions, one can imagine that mastering proper sequence is one central difficulty in the making of *zisha* pots.

Gongfu

The fact that Ge was capable of working without pausing to think about his next move indicates a certain disciplining of his own body. Control of movement is referred to by the term *gongfu*, the same term indicating the martial art kung-fu (another transliteration of the same word), often translated as 'precision in movement'. Mastering the craft involves mastering each of the movements involved in the making process.

The term might conjure images of highly skilled martial arts practitioners, but in *zisha*, *gongfu* refers also to some of the less impressive and apparently easier techniques. An example: making slip (wet clay used to combine elements) requires the crushing of dry clay in water with a bamboo stick against a slab of stone. To get the clay as finely crushed as possible requires time and the proper use of the tool. A series of movements are used: taking a little of the clay, pressing it against the stone, going through the clay and, with a circular movement, bringing the clay together before repeating the process. When I was learning this way of making slip, an artisan who had walked into the workshop told me 'this also is *gongfu*', in other words this simple technique is also something that has to be mastered – and that can have consequences for the final object. Indeed, clay that was not crushed finely enough will be apparent on the otherwise smooth surface of the pot.

I have mentioned the tedious process of learning to properly perform movements at the early stage of apprenticeship through endless repetition. However, this does not mean that *gongfu* is to be understood only as 'correct' movement. Some artisans are said to have better *gongfu* than others, but it is not only a matter of degree of skill; artisans also recognize that *gongfu* can be developed in personal ways. For instance, I heard artisans discuss the *gongfu* of another, said to have very good *gongfu*, and who would beat the clay particularly hard, with exaggerated movements. They said this was not the way they did things, but it seemed to work for that particular person. So although *gongfu* is initially constrained, literally through the actions of the teacher who forces the hands of the student into place, it can also be developed creatively into a personalized skill. There is an interesting blurring between correctness and creativity in *gongfu* – or more accurately, one does not find in this case the division created in Europe in the seventeenth century between creativity and repetition, freedom and determination, and ultimately art and craft (Coleman 1988:7; Ingold 2000:350), a point I return to in the next chapter.

Traditional and modern tools

I became quite interested in the tools potters used in Dingshu. Firstly, one cannot but be surprised at the number of different tools necessary to shape an (entirely handmade) teapot. Unlike other pottery traditions, the bare hands do not directly shape the clay, save for a couple of marginal processes. This means that at every stage, and for every different curve and shape, a different instrument needs to be designed, made and used. Many of the tools employed are, subjectively speaking, extremely beautiful. For those who can afford them, some tools are made from hard woods such as Chinese mahogany. Bamboo tools are sandpapered until perfectly smooth, and years of handling give them a particular sheen. Artisans also have a certain fascination with the tools they use or make, and as they visit the workshop of a colleague they will at times handle and admire new tools acquired or made.

The same perfectionism that goes into the making of pots also goes into making tools That is because, I was told, without perfect tools one could not make perfect pots. Those artists who make pots by hand – indicating that they are traditionalists – will have in a corner of their workshop or house a workbench where they can make the tools that are necessary for the new designs they create. A married couple of artisans told me that, between them, they had made over 2,000 tools of all shapes and sizes over their careers. In order to make these high-quality tools, artisans need to master tools (adze, saw, various types of files), which have themselves to be of suitable quality. My informants were categorical about this: German tools are the best, Chinese tools are of no use. This illustrates the care that goes into the making of tools and their importance, since German tools are not easy to find nor cheap.

I have mentioned that in the old factory all artisans made their own tools. Today however, some artisans complain that the skills linked to the making of tools have almost disappeared. It is now possible to buy tools from one of the three shops that have appeared in recent years opposite the buildings of the Zisha Factory. These tools are machine made and of poor quality, and have to be modified by the artisan to be of any use – for instance by sand-papering rough edges. Many artisans told me that they simply did not buy such tools, because they did not meet the high standards needed for their work. They might rely instead on one of a few artisans who have specialized in the making of handmade tools, which they sell at cost. Mr Pan is such an artisan. He recovers wood from old furniture, in particular Chinese mahogany, which is ideal in that the wood, several decades old, will distort very little with variations in temperature or humidity. Pan's tools are also very beautiful, and I have seen artisans look at them admiringly, commenting on their beauty and feeling their smooth surfaces with their fingers.

I have mentioned that many artisans claim that *zisha* is a dying craft, and that many of the techniques and skills are being lost. The main cause for concern, according to them, is the introduction of certain modern tools. These tools change the relationship between the body of the artisan and the craft object being formed. One of these modern tools is the metal hand-turned wheel. When artisans work on their pot, they need to be able to turn it while working on shaping the sides or smoothing the surface. The traditional tool used is a simple wooden disc, rounded at the base (so it can turn) and flat on top. To make it turn, the left hand grabs both the pot and the disc together, whilst the right hand stays in place, for instance pressing a bamboo tool on the pot's surface in order to shape or smooth it. The metal wheel (which a European potter would recognize) is more sophisticated. It consists in two flat metal discs joined together by a central column – one disc rests on the table whilst the other turns on the axis. This wheel is operated in very much the same way as the wooden disc, by the left hand, but it has clear advantages. First of all, it is much more stable than the wooden disc. Its weight keeps it in one place on the table, in contrast to the lighter wooden disc, which has a tendency to wobble and to move across the table if not controlled properly. The second advantage of the metal wheel is its inertia. This is especially useful when shaping the body of a pot by patting (*da shentong*) because the pot turns more evenly and it is possible to achieve greater regularity. Thirdly, for the reasons mentioned, an apprentice spends less time learning to control the metal wheel. For these reasons, the vast majority of artisans use one or more metal wheels, but also use one or two wooden discs for certain easier operations.

The traditionalists who continue to use the wooden disc, however, hold that the metal wheel disrupts central principles of the craft. Given mastery over it, according to them, the wooden disc becomes an extension of the body, and responds better to the movements of the artisan. The reason is that the wooden disc is only a few centimetres high, against about 12 centimetres for the metal wheel. This changes significantly the way the artisan works: with the wooden disc, the elbows can rest on the table, and it is the body that gives stability to the tool. In contrast, with the metal wheel the body has less stability and cannot control the process as efficiently.

Other tools follow a similar rule. The *bizi*, the tool used to perfect the shape of pots after rough shaping by patting, giving the pot its profile, is a case in point. The line of the *bizi* will determine the line of different parts of the pot. This means that a different set of *bizi* will be required for different designs of teapots, and also, in the case of some masters and for some pots, a new tool will be made for an individual pot. The *bizi* is made out of a piece of the side of a section of bamboo. One can appreciate the use of bamboo here: because teapots are usually rounded, bamboo is ideal since it is already curved, and is only to be slightly modified to take the shape one wants to give a pot. The other advantages in using bamboo are that it can be cut in a perfectly straight line when cut in the direction of the grain – this is particularly useful for this and other kinds of tools which are required to be straight. It is also extremely solid, and if old enough – the bamboo should be used at least three years after it has been harvested – it will not change shape. Also, after it has been filed and polished it is extremely smooth – the *bizi* is not only used to shape a teapot, it is also essential in the first stages of smoothing the surface of the pot. In fact, it might be safe to say that without bamboo, pots would be made in an entirely different way in Dingshu. When making the *bizi*, the part of the bamboo that will be in contact with the clay is filed in the direction in which the tool will be used, not straight out, but at an angle. This, my teacher Zhao tells me, is how Gu Jingzhou taught him to make tools in general: he would always think about the way in which the tool would be used, and how the hand would best grasp the tool. Filing the *bizi* in such a way means that when the tool is held, it becomes an extension of the hand, and the tool will smoothly glide on the surface

With the use of the plaster mould, introduced in the 1950s, such a tool is needed, but it does not serve the crucial function it does with handmade pots. Because the profile of the pot is created by the mould, the *bizi* is only used to smooth the surface of the clay. In fact, the *bizi* is often replaced by a simpler tool, a flat piece of plastic or wood in which is cut out the line to be given to a pot. This tool is held perpendicularly in the right hand, whilst the left hand turns the wheel. Here also, the relationship between hand, tool and clay has changed. My teacher Zhao pointed out to me the differences: with the traditional *bizi*, all the fingers can grasp the tool, instead of the tip of three fingers that are used with its substitute, so that the tool is much more stable and pressure can be better controlled. The *bizi* is an extension of the hand in that it enables the artisan to feel and respond to pressure on the surface of the clay more efficiently than other tools.

More than a statement of fact, the claim that modern tools change the relationship between the artisan's body and the clay feeds into the discourse of the exclusivity of the craft that I mentioned in previous chapters, and which I discuss in more detail in the next chapter.

Conclusion

The process of learning skills in Dingshu relies on repetition, a certain discipline of the body, and instruction, more or less explicit, by the master. In that sense, it resembles many other types of craft. But by being sensitive to the elements that were stressed in the discourse of artisans, I have highlighted some of the particularities and cultural salience of the craft. One important point to stress is that artisans talk of skills as being developed in their own right; they become more than a means to an end, and in the quest for a precision of movement, or *gongfu*, they are ultimately almost reified.

The idea that a body of knowledge, a tradition, gets passed on in one block from one generation of artisans to another has often been used in the literature as a useful short cut, but it was necessary to problematize it in this chapter. Rather than a text which one might edit, remove parts or whole paragraphs of or add corrections to, a craft can be said to be reinvented anew for every individual through the development of skills. This reinvention takes place within a framework of techniques which are passed on from generation to generation not in a static manner; rather, they are perfected throughout the artisan's career. This dynamic relation to skills is directly related to their incorporation: as they become part of one's body, of one's person, they take on unique characteristics. It is a short step from saying that skills and tools become part of the body to saying they become part of an identity; this will be the theme of the next chapter.

CHAPTER 6

Style, names and authenticity

Zhao Jianghua was the last student of the master Gu Jingzhou, having taken up the craft not as an apprentice recruited by the Zisha Factory, but after he started looking after the health of the ageing master as one of the factory's doctors. Zhao started learning the craft not out of self-interest but after a request from the master to prepare bamboo tools for him – the master complained of a lack of strength in cutting this particularly hard material. Later, Gu encouraged Zhao to take up the craft, which he did, eventually giving up practice of medicine to devote himself to making *zisha* pots. This close relationship with his late master went side by side with a close relationship with the objects the master produced. Although he refused to consider himself a particularly skilled artisan, Zhao would explain that he had developed a profound knowledge of the work of his master after spending many years in the same workshop, his worktable facing that of his master. This knowledge of his master's work was such that Zhao maintained that he could recognize at a glance whether a pot had been made by his former master. This capacity had a practical use for him: he and some other former students of the master were authorities on determining the authenticity of items said to have been made by Gu. Collectors visited Zhao and his colleagues in Dingshu in order to receive their expert, and authoritative, advice.

This chapter brings together several themes related to the topic of personal style in *zisha* pottery, a topic that would often be raised in my conversations with artisans. I start this chapter with a note on the relationship between a master and his former doctor and student, and the capacity of recognizing objects as having been made by a particular artisan, as an illustration of the close relationship, and occasional overlapping, between people and things in Dingshu, a theme I have alluded to so far. This story also raises a conundrum: as I discussed in the previous chapter, the precision and perfectionism in realization, the care put into the making of tools, suggest a craft that strives towards the precision of engineering. When such precision is put into the craft, how then is one to understand the claim that pots bear marks of their maker? I use a discussion of the term *fengge* ('style') as a point of entry to address this apparent conundrum. I continue discussing the topic of the relationship between artisans' bodies, materials and tools introduced in the previous chapter, but take this issue further by looking at the way this relationship is caught up in discursive practices that arise as a result of the turn to capitalism in Dingshu: competition, fame, new forms of division of labour, authenticity and legitimacy. Some theorizing in the social-science literature on the agent-like attributes of things completes the analysis.

The two previous chapters have been concerned with the development of skills – their acquisition by novices in Chapter 4 and development by artisans throughout their careers in Chapter 5. One might characterize the development of skills as a gradual process of gaining freedom, freedom from the scaffolded control of the master, as novices more directly engage with tools and materials. In the exchanges between peers discussed in Chapter 4, new techniques and new design ideas are gleaned in exchanges with peers and assembled to create designs and develop techniques that were not expressly taught by a master. Greater freedom is still gained once artisans have mastered the basic techniques of the craft and can explore personal ways of working with the clay, as with the artisan introduced in the previous chapter who was known to beat the clay more vigorously than his colleagues. This gradual increase in an artisan's freedom will eventually lead, at least according to artisans themselves, to the development of a personal style, *fengge*. Understanding the meaning and implications of *fengge* drove my ethnographic investigation. On my first meeting with an artisan, a question I would often ask was: What is your personal style, your *fengge*? This, I felt, was an easy entry point into an interview, giving an opportunity for artisans to talk about their work and what, they thought, was special about it. But a remark on *fengge* would also come up spontaneously in my discussions with artisans.

This chapter follows this process of investigation, following my understanding of the term as my fieldwork developed, and subsequently how I made certain links between phenomena that were apparently not connected. We will see that *fengge*, as with the corresponding English term 'style,' can refer to the style of a person, of a category of the population, of nations or of a period of history. In the second part of the chapter, I change my approach, and reflect on the contexts within which the term *fengge* is used. Of particular interest here are contexts of determining authenticity, which define the *fengge* of individual artisans. Thus, *fengge* has deep implications in the art world of *zisha* ceramics, in relations with fellow artisans, collectors, connoisseurs and museums.

In the simplest terms, *fengge* can be defined as that corpus of 'signs' that identify the maker of a pot. *Fengge* is a word of common usage in Chinese, and holds similar meaning to the English word 'style.' In the same way as in English, one can speak of the *fengge* of a certain period of art history, of Occidental and Oriental *fengge*, and the *fengge* of groups and individuals. As one Westerner art theorist puts it, 'connoisseurs of the arts have no difficulty identifying a Mozart or Haydn quartet, a Keats or Shelley sonnet, a Braque or Picasso painting, even when they have never encountered the particular work in question' (Gardner 1971:515). Style is a short cut for describing a phenomena that is elusive, and that historians of art and anthropologists have not agreed upon, or have decided they should not (Arnheim 1981). Style is elusive because it escapes proper definition: any element of the work of art might constitute a style. The art historian Schapiro proposes a definition: 'by style is meant the constant form – and sometimes the constant elements, qualities and expression – in the art of an individual or a group' (Schapiro 1953:287). It would be difficult to find a more general definition: style is a question of both form and quality, recognizable sometimes from elements, sometimes from the whole. Yet a useful working definition – if one needs one at all – can only remain broad in scope.

My intention here is not to discuss the style of individual artisans, nor of *zisha* craft as a whole. Rather, I single out the notion of *fengge* as particularly useful in my broader discussion on the relationship between people and the objects they produce in Dingshu. If one accepts that it is possible to identify the artisan by looking at an object, then a few preliminary questions need to be addressed. What is this mechanism by which signs of the artisan's individuality are impressed in the clay? And what, precisely, is so transferred and in what form? At a first level of analysis, and according to the way artisans talk about *fengge*, for instance when authenticating a master's work, *fengge* points to the idea that the engagement between the body and tools and materials I discussed in the previous chapter results in a strong link of identity between artisan and creation. At a second level however, and in a more critical stance, I want to show that this relationship does not arise spontaneously but is constructed through discourse and practice. To do this, rather than taking the notion of *fengge* literally, I will deconstruct the process that leads to the infusing of a pot with the signs of its maker. The main paths along which the link between author and object develops and manifests itself are the practice of stamping a name on the pot and related discourses about the authenticity of individual works. The discussion will accordingly unfold around these issues.

Is *fengge* personality?

I have mentioned that, in the same way as 'style' in the English language, *fengge* can have many different meanings, and its exact definition is elusive. In Dingshu I received many different answers as to what it might be. One contradictory answer had to do with its link with an artisan's character. In one of my first interviews in Dingshu, one artisan, Chen, tried to explain how, having a good knowledge of *zisha*, one can figure out from the line of the pot many features of the identity and personal qualities of the person who made it. These included whether the maker was a man or a woman,[1] and whether he or she had been brought up in the countryside or in the town. I would later hear many times this statement that works made in the villages have a different *fengge* to those made in town. I was told that villagers make pots that a 'rougher' (*cu*) in appearance. This was not always a judgement on their quality, but simply an assessment of their characteristics, the lines of the villagers' pots tending to be fuller and bolder than those of townspeople.[2] Chen herself spoke of her childhood in a village near Dingshu, and explained that choosing to become an artisan was for her a way of escaping the difficulties of rural life. Chen used natural motifs extensively in her work, and linked those to her love of observing nature in the countryside.

Soon after that first encounter with the idea of style and the character that is imprinted on pots, I asked another artisan, Mr Fang, who was to become one of my main informants, whether his pots revealed his personality. 'Yes,' he said, 'they are like me, frank and straightforward.' Mr Fang was referring to the plain forms of his usually square pots, as well as his reputation of not being scared to hurt other people's feelings with a frank appreciation of their work. Without prompting, he used the term *fengge* to talk of this similarity between personal character and object. However, in further discussions with the same artisan, I came to realize that *fengge* was a concept that was used in a way that implied much more than character. According to Fang, not anyone can be said to have *fengge*; rather, an artisan develops *fengge* after at least twenty years of hard work, although many do not manage to develop *fengge* at all. Fang was here linking *fengge* with technical excellency.

However, further vagueness surrounded the elements that made up *fengge*. When I tried asking what elements constituted *fengge*, I got a range of answers. At times, it was the subject matter that characterized *fengge*, such as

---

1 Interestingly, in her study of calligraphy, Yen (2005:68) reports that her informants also spoke of their ability to determine whether the originator of a piece of handwriting was a man or woman.

2 This is another way in which villages and town are set apart, as discussed in Chapter 2.

in the case of Chen. In another case, an artisan, Yao, told me of his *fengge* in terms of elements that appeared on the surface of his work. He told me of the simplicity and fullness of shapes that characterized his pots. Another artisan identified the simple fact that his works were plain and unadorned (*pusu*) in form, and this also characterized the *fengge* of other artisan members of his family. Perhaps this artisan was talking of a different meaning of the term *fengge*, the conscious decision of working on a certain style of pot, rather than the more elusive 'personal style' that is said to characterize the works of an accomplished artisan. This shows that the term, as is the word 'style' in English, remains ambiguous. When I asked my questions, I felt I was being interpreted in very different ways; the answer might have been the one that suited the situation rather than an attempt at an identification of *fengge*. For instance, the artisan who told me about the *fengge* of his family, in particular their father and grandfather, might be trying to promote his work with reference to his more famous family members. Yet another of my informants, Wan, even told me he believed there was no such thing as the *fengge* of an individual, but only of periods in the history of *zisha*.

Why then was the term *fengge* such a useful one to use for artisans? Another point Fang made about *fengge* is helpful to solve some of its contradictions. Fang claimed that it was up to connoisseurs or masters to define the *fengge* of other people: individuals cannot themselves claim to have *fengge*, and even less describe what this *fengge* might consist of. In that sense, Fang was restricting the possibility of having *fengge* to only those artisans who become talked about in the specialized literature and talented enough in the eyes of fellow artisans. *Fengge* can then be in an important way connected to fame. This is the perspective I take, because it enables a more critical appraisal of what *fengge* is. *Fengge*, in that sense, is also closely linked to the practice in *zisha* pottery of stamping a personal name on the base of a pot, which I discuss later. I start however with a clarification. One might object that there is an important difference between a painting by Picasso and a *zisha* pot, namely that one is art and the other craft, and that as a result the notion of the *fengge* of ceramic objects is very different to a Western understanding of artistic style. Yet what precisely is intriguing with the notion of *fengge* as applied to a craft is that it shows that the distinctions between art and craft, intellectual and manual labour, creativity and repetition, freedom and determination that have characterized Western thought since the seventeenth century (Ingold 2000:350) do not apply to *zisha* craft, or indeed, as Ledderose (1999) argues, to Chinese arts and crafts in general. The distinction between such 'arts' as painting and calligraphy on the one hand, and 'crafts' such as lacquer production, sculpture and ceramics on the other, rests on rather different principles, which I now turn to.

Art, craft and factory art

In a study of Chinese art in an age of mass production, Ledderose (1999) reminds us that much of what we think of as Chinese art is rather what he calls 'factory art', commercial objects produced in the context of a factory, with control over quality and management resting with individuals who might not be artisans themselves. He includes in this category sculpture, lacquer work, bronze, silk and ceramics. To these, he opposes the two main Chinese 'high arts', calligraphy and painting. By 'factory art' Ledderose means something different than the type of factory that appears in the context of industrialization. The industrial revolution in China did not take place until the twentieth century, though most of the necessary elements of such a revolution had been present for centuries, such as efficient specialization, the division of labour and management practices, and modular production systems. 'Factory art', for Ledderose, involves systems developed to optimize the production of craft objects, systems that render work repetitive but allow for great flexibility. In his view, China became the pioneer of modular systems that, although standardizing the process, enabled the production of a great diversity of forms with cost and time efficiency. An example he gives is the famous army of terracotta warriors in Xi'an. It is usually said that every single statue is unique, but in fact, the makers of this army resorted to an ingenious system whereby different parts of the bodies of the soldiers were made separately (often by different production units) and serially, and designed to interlock. These parts were then combined in different ways to create the illusion of uniqueness. For instance, the same face would have been used many times, but each time combined with a different shape of head.

Calligraphy and painting, the 'high arts' of China, employ a similar approach. In painting, for instance, elements such as bamboo leaves or flowers are at first learnt individually by the novice, and at later stages of the learning process are combined to make a composition. Some Chinese painters were known to only make paintings of the same subject. For instance, a painter might only paint bamboo, but vary infinitely the display and composition. Yet, for Ledderose, there is a fundamental difference between these arts and other modular crafts:

> Whereas craftsmen have been aiming for ever greater standardization, and have achieved ever more perfect conformity of modules and units, calligraphers have been doing the opposite. They consciously exploit variants that creep in unconsciously in the execution of characters, trying new shapes, finding new forms, studying previous masters' inventions, modifying their creations and interpreting them. Generation after

> generation, countless practitioners built the increasingly complex edifice that is the great calligraphic tradition in China.
>
> (ibid.:194)

Ledderose is here putting forward two ideal types. Both rest on the principle of modularity, but in one case, that of craft, standardization occludes the signs of the artisan's individuality. In the case of painting and calligraphy, by contrast, individuality is made apparent despite standardization. One could say that craft sublimates the relation between maker and product, whereas art reveals it. In that view, creativity is enabled precisely through repetition. Calligraphers and painters need to master perfectly, almost mechanically, the written characters or elements of a painting until they become second nature; when this is achieved, the minute variations of the characters and paint elements caused by spontaneity appear and make the work unique. This understanding of the nature of art turns on its head the equivalent Western distinction between art and craft whereby spontaneity cannot be achieved as a result of mechanical repetition (see Ingold 2000:350).

Ledderose argues that the fundamental difference between craft and art in China is the possibility of spontaneity. He talks of calligraphers as developing spontaneity by transcending the mechanical repetition of characters. Can an artisan be spontaneous working with clay? It is here that the difference is made apparent. The 'freedom of the brush' is precisely limited to calligraphy and painting. If something of the identity of the artisan transpires in the work, it is through a process different from – and I would say, quite opposite to – spontaneity. As such, the idea of modularity fits the case of *zisha*. It describes well the process of making a pot, which is built from different elements cut out of sheets of clay – body, base, neck, handle – and maybe also the process of creating a new design: finding the spout and handle that will fit on a particular design of body. On the other hand, it is also true that some artisans are engaged in an activity similar to what Ledderose describes for calligraphy and painting: they are reproducing over and over the same design, inspired by previous masters' renderings of it, but varying some of the elements (in particular, the line of the body) to reach the most satisfying materialization of the design. Significantly, these artisans identify their work as 'works of arts and craft' (*yishupin*), as opposed to what they call commercial works (*shangpin*).[3] Should one then conclude that *zisha* can be compared to calligraphy? To answer this question, one needs to have a closer look at the

3 The dichotomy usually also distinguishes works made by hand and those made using a mould.

zisha making process. If calligraphers achieve singularity through repetition, what is happening in the case of *zisha*?

Designs and making

How is one to understand the claim that *fengge* takes a minimum of twenty years to develop? One might say that in a similar way to calligraphers, artisans need to find their own 'freedom of the brush', which implies a perfect mastery of technique (as discussed in the previous chapter in terms of *gongfu*). I would complement this explanation by trying to conceptualize the process of overcoming techniques. I have mentioned that apprentices learn each stage of the making of a pot in succession, preparing the sheets and discs of clay, patting the body into shape and adding the different elements, over their three-year apprenticeship. This has the effect of creating continuity with the apprentice's later career. In the artisan's daily life, the three years that were employed to learn the sequence will be compressed into a few hours or a few days, the time necessary for crafting a pot from start to finish. Throughout their entire career, the strict sequence I have mentioned is reproduced from beginning to end, and started over again. This might sound tedious, but many informants actually talk of the pleasantness of the process. When she is working, Chen told me, she forgets about time, about worries. Other artisans, however, are not as positive as Chen, and complain of the long, boring hours spent working; in many cases, a radio or miniature television placed on the worktable help alleviate the boredom.

As I mentioned, *zisha* artisans are very aware of the fact that a completed pot is the sum of the gestures that went into the making of it, an accumulation of stages done more or less well (one has more or less good *gongfu*). If the initial stages have been badly realized, the finished product will show it, which is why a master will never use the discs and sheets of clay prepared by an apprentice, even a talented one. It is through this cycle of making that an artisan can learn to perfect a design. I have already referred to the collaboration between Wan and an art professor from Beijing. Wan told me he had to correct, over several trials, the 'mistakes' of the design. The design of a *zisha* pot can be perfect on paper, precise to the millimetre, but the concretization of the design can never be the exact reproduction of what is drawn on paper: it is only when the design is embodied in a pot that one knows whether the design 'works' or not. Wan was adamant that people who have never made *zisha* cannot know how to design it properly. The pot therefore goes through different phases. Each phase is fairly lengthy: making a pot might take a week, plus two days before the pot is returned from firing. The artist will look at the finished product and decide what its flaws are before starting over with another pot, attempting to correct the mistakes. Wan told me that it usually took him five tries before perfecting the design. For each new try, he has to translate the changes into movements. When patting the body of the pot, the fingers must be angled in a slightly different way, the sequence must be changed slightly. If the changes are to be made in the spout, an artisan will have a better chance of getting things right, as they can make several different spouts at once and decide in the final stages of the work which works best. Zhao Jianghua told me that his master Gu Jingzhou once spent eighteen days making spouts before he got to the exact shape he wanted – another example of the perfectionism that can be involved in the making of *zisha*. Of course, being able each time to make a pot (or spout) that is better than the previous one, or that incorporates envisioned changes, is a feat that only the very talented are able to achieve. For the others, making a particularly good pot is in part a question of chance – of choosing the best out of many trials. This might start explaining why *fengge* takes so long to develop: a high degree of skill is needed to reach the stage whereby each successive realization can incorporate those changes willed by the artisan.

This cycle would sound familiar to anyone engaged in any art or craft. It has been theorized by psychologists and philosophers working on the idea of creativity. Götz (1981) cites the psychologist Catherine Patrick, who proposed a systematic account of the process of creativity, which according to her had four clearly defined stages: preparations, incubation, insight (or illumination), and verification (concretization). Götz however identifies creativity with the stage Patrick calls verification, and adds a further two stages that are to him essential in this cycle of creativity: product, and evaluation (in terms of morality, usefulness, originality, beauty and so on). Götz's argument is that creativity cannot be known until realized; its potential can only be known after its actualization. Insights are moments, but creation is a process; psychologists say they are studying the creative process, but what they are analysing is only, in Götz's argument, the stage of incubation. But this incubation can only be known a posteriori, from its concretization.

The two further stages Götz puts forward are crucial in an analysis of the creative process, as they are the moments at which improvement can be made from the perceived mistakes that appear in the initial concretization of ideas. This is the classic case of objectification, which is defined, in Miller's reading of Hegel, as:

> a dual process by means of which a subject externalizes itself in a creative act of differentiation, and in turn reappropriates this externalization through an act which Hegel terms sublation (*aufhebing*). This act eliminates the separation of the subject from its creation but does not eliminate this

> creation itself; instead, the creation is used to enrich and develop the subject, which then transcends its earlier state.
>
> (Miller 1987:28)

Artisans in Dingshu put these points across most forcefully when they say that a design, even perfectly drawn, must be worked on at the stage of making. It is common-sense to Wan and other artisans that creation goes through such cycles. Ideas cannot simply be translated into clay form – it is through the process of creation, the interaction between clay, hands and intellect, that ideas are materialized in a pot.

What happens in the new embodiment of a design? An artisan will look at their previous work and decide what its flaws are. These might be verbalized according to certain conventions, such as the three-tier division of the body of the pot into neck, shoulder and belly. The artist might think the pot is too fat because the belly sticks out too much, or there might be an imbalance between spout and handle. Wan once showed me two pots he had made, a first and a second version of a pot, to illustrate the different stages he went through before finding the best embodiment of a design. The differences were easily distinguished; the middle part of the pot, for instance, came out more abruptly in the first version, and the distance between spout and lid was longer than that between handle and lid. Intriguingly, Wan talked about these differences in terms of correctness: the second version was more 'correct' (*dui*) than the first, in his view, because it is more 'beautiful' (*haokan*). How does Wan manage to conflate beauty and correctness? I would suggest that, because the different parts of the pot are understood in terms of parts of the body, the evaluation process itself is one that is embodied.

I mentioned that pots are described in terms of parts of the body in order to verbalize one's evaluation of it. The aesthetic criteria for the evaluation of a pot follow those of the body: the pot should not be too fat though suitably plump; the central part should not protrude too much, there should be a certain fullness in the pot, like the chubby but taut skin of a baby. Artisans might comment on the knob on the lid being too large, for instance, by saying its hat is too big, and give a short tap on their own head. One artisan tried to illustrate for me the principles of elegance in a pot by pretending to be a teapot himself, one arm acting as the spout and the other as the handle (as in a nursery rhyme dance), and taking on different airs to suggest different types of pot. I would suggest the body is both used to think about, evaluate, the shape of the pot, and, as a corollary, as an aesthetic canon. A pot that resembles what is thought of as a beautiful body is a beautiful pot. Or probably more accurately, as the example of the chubby baby implies, the image of the healthy body makes the pot beautiful.

It is interesting to note that anthropologist Yen Yuehping (2005:76) makes the same remark in her study of Chinese calligraphy. Yen refers to Gell's concept of the extended person to explain Chinese people's attitude towards calligraphy and handwriting. She quotes informants who suggest that the handwriting of a person reflects the person's inner dispositions, moral character and so on. One might recognize this as similar to the Western practice of graphology, though Yen suggests an important difference between graphology and Chinese calligraphy. In graphology, the unique characteristics of an individual's handwriting are understood as a natural organic movement, the result of sensory motor dispositions and patterns, which betray the inner dispositions of the mind. 'In other words', Yen explains, 'in the tradition of Euro-American graphology, the relationship between handwriting and personality is firmly connected to the idea of a mindless body that nevertheless reveals the secrets of the mind' (ibid.:66). In contrast, the whole body is involved in Chinese calligraphy, and what is more, the body and calligraphy are thought of as isomorphic, so that a well written character is one that reflects the image of the beautiful and healthy body: 'the Chinese believe that the person and the handwriting are mutually generative' (ibid.:66). In graphology, hand tremor is seen as betraying the inner dispositions of the person. Chinese calligraphy by contrast involves both body and mind, and Yen suggests the use of the term 'body-person' to describe the 'extended person' of Chinese calligraphy (ibid.:75–80). I mentioned in the previous chapter that different artisans are said to have different *gongfu* – that is, the way they have developed their skills varies – and I have now talked of the relation between body and object, and the process through which the concretization of a (body) form is evaluated in the *zisha* cycle of creativity. *Fengge* can then be seen as the extended body-person of the artisan.

Yen talks of the moral characteristics that transpire in handwriting, but I have not heard artisans talking of *fengge* in such terms – save for one possible exception, addressed in the next chapter in the context of mastery. My case follows that logic, but is a little different. I have suggested that it is not character that is inferred from the signs of *fengge*. What, then, does *fengge* 'do'? I would suggest, simply but crucially, that it identifies the person who made an object. Purchasers are not particularly interested in the qualitative elements of *fengge* when buying a pot – in other words, they are not interested in the aesthetic appreciation of these signs, but rather that the *fengge* confirms the identity of the maker; they want to know whether their prized possession is genuine or not. To understand the implications of this idea, and to understand the relation between *fengge* and identity, I turn to considering the importance and significance of names in *zisha* pottery.

Names

A specialist on the history of *zisha* pottery, Bartholomew (1990) has suggested that the act of inscribing a name on a pot was an indication that the makers considered themselves to be true artists, at the same level as calligraphers and painters. I am, however, not convinced by this interpretation. First of all, there are many cases of Chinese crafts where a name is carved or stamped on an object, but where the purpose is clearly one of quality control. Ledderose (1998:78), talking about the production of lacquer in ancient China, mentions a lacquer beaker upon which are listed no less than seven makers and the names of four supervisors. Why would it not be so in the case of *zisha* pottery? Historically, there was an important difference: the crafts Ledderose is talking about were destined for the Imperial court. The market for *zisha* pottery by contrast was driven by the class of literati, who were familiar with collecting calligraphy and paintings; indeed, they produced them. The seals that adorn *zisha* pots (Figure 16) arguably bear a closer resemblance to the seals affixed to calligraphic scrolls than seals meant for quality control. That said, that artisans were claiming that their craft is equal to calligraphy or painting is doubtful. For the literati of the Ming and Qing, the superiority of painting and calligraphy was beyond discussion.

I cannot pronounce on the original reasons for stamping names on pots, but I argue for the continued relevance of its use today. The main reason for signing pots is that buyers expect to find such a seal stamped on their purchases. It is common to see purchasers turn over a pot to glance at the stamp situated on its underside, even if most of the time the seal will not say much about who made the pot: given the numbers of mass-produced pots made throughout the region, the name will not mean much to the purchaser. In fact, the seal might not even be decipherable, since names on seals are often written in an archaic script. It might not even contain the name of the maker but a nickname or aphorism (which come to be the trademark of artisans). However, the seal has become a kind of trademark of *zisha* craft, so its absence would be an indication that a pot might be of different origin. For more 'upmarket' *zisha* pots, the name has further importance: it is an essential element in the economy of fame into which artisans enter at the start of their career.

Becoming famous, or 'having a name', (the literal translation of the Chinese term for 'fame', *youming*) depends on a number of strategies that artisans rely on. Except for the cheapest (and interestingly, also the most expensive), pots are usually sold with a certificate of origin. This certificate takes different forms, either a calligraphed note describing the pot and stamped with the seal of the artisan (the same as the one stamped on the pot), or a printed certificate which usually includes a short c.v. of the artisan, a name seal and a photograph of the artisan holding the pot that is being purchased. (During more recent visits

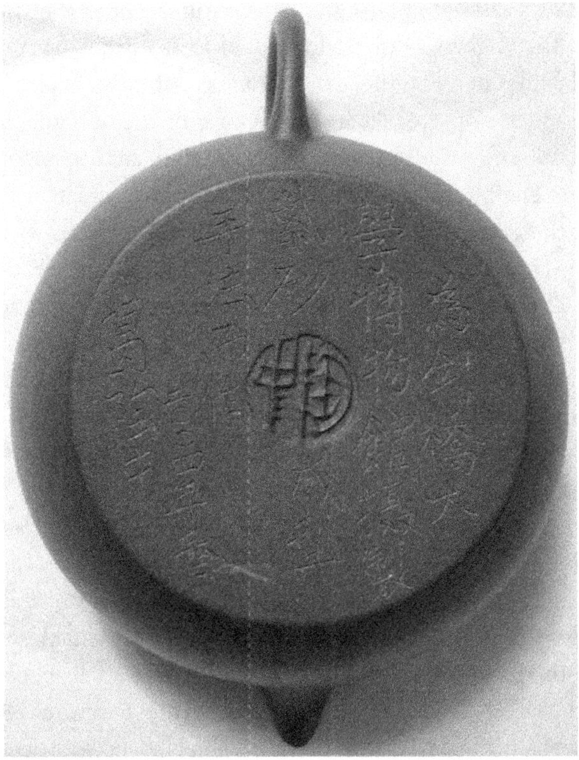

Figure 16 *Seal at the base of a teapot made by Ge Taozhong. The artisan also added an inscription at the base, a common practice when pots are commissioned (in this case, the pot was commissioned by the author on behalf of the Museum of Archaeology and Anthropology, University of Cambridge).*

to Dingshu, I was told that the practice of including such a photograph was outmoded.) Often, a genealogy indicating how the artisan is related to a master is also included, and in these cases the photo included might show the artisan and master side by side. The purpose of the certificate is twofold: it establishes the link between person and object, and circulates a name, face and biography with the pot. This is not only important for the artisan, but also for the buyer; I was told that the certificate and the photograph are essential to the buyer, not so much as proof of authenticity but to demonstrate the link between the object purchased, the artisan, a place (Dingshu) and probably a master (if included in the photograph). In fact, many purchasers also ask for a photograph of themselves standing next to the artisan, with the pot being held by either individual. The archetypical photograph included in the certificate shows the artisan holding the pot in their upturned palm, and either looking straight into the lens of the camera or, more dramatically, studying the lines of the pot.

The pose artisans assume in photographs is reminiscent of the way an artisan would evaluate the lines of a pot during the process of making it. At the stage of stabilizing the lines of a pot, the artisan must constantly check that the shape is satisfactory, and if not, correct it. In the case of the vast majority of pots, this stage does not enter the sequence of production, as the shape is stabilized with the mould. The photograph might then be said to create a further short cut: looking at the pot in this way might indicate that the pot was handmade. Other strategies for establishing fame are participating in competitions (several of which are organized by the county government) and accumulating awards. Awards and other certificates are invariably displayed on the walls of workshops or in showcases. But the ultimate aim is to be included in one of the many books on 'famous artisans' published in Taiwan and Hong Kong.

At the very least, the fact that *zisha* pots contain the name of their creator means that they are meant to be circulated. Nancy Munn (1992) has also described fame on the island of Gawa in Papua New Guinea as the association between valuable objects and personal names. Gawans say that one can know a person by face and by name, and fame is precisely this situation where the name is known but not the face. In the *kula* exchanges of shell valuables that Gawans engage in with the inhabitants of neighbouring islands, shells become vehicles for names, and as such they become, in a way, spatio-temporal expansions of the self (ibid.:117). In the same way as *kula* valuables, *zisha* pots are vehicles for the name of their creator, but this time the name is physically inscribed on the pot. Names are not part of speech as in the circulation of *kula* shells, they are instead made material. I once asked an artisan, Wan, why he thought his works were valued so much, how he gained fame. He suggested that collectors might have bought some of his works along with other pots by other artisans in the 1980s. At that time, overwhelming demand for *zisha* teapots from Taiwan meant that some pots were often purchased before they were even fired. Wan imagined that, after a few years of use, his pots had stood out from the others in their beauty. Wan was referring to the fact that, as they are used for drinking tea, teapots develop a sheen, which is more apparent when the pot is made of quality clay and of superior workmanship. According to Wan, these same collectors, having looked at the name on the pot, returned later to purchase more of them. So the inscribed name is crucial to permit this identification, as it is in cases where connoisseurs decide to write an article about an object in their collection, which in turn will contribute to the fame of the maker.

Bosses, employees, forgeries

Of course, a name is not necessarily a proof of identity of the maker. I have mentioned previously the practice of *jiagong*, whereby a 'boss' (*laoban*) employs a 'worker' (*gongshou*) to produce pots of their design that are then stamped with the boss's name. I once visited a young artisan who worked from his apartment in a building near the town centre. He was making a version of a famous pot by Shi Dabin, the sixteenth-century master. For this, he was using a plaster mould that another had had made. The artisan had stamped a name on the bottom of the pot, and when asked whose name was on the seal, he answered it was the name of the person who had designed the mould. The artisan said the pots were sold for about 80 yuan. He only made about six a day (as opposed to the twenty a day some people manage) because much care was put into the finishing touches. For instance, the joints created by the moulding process had to be carefully concealed. Because the design was a little complex (it was to resemble a seal wrapped in cloth), the craftsman had also to redo some of the lines mimicking the ripples of cloth. Although the whole design was already contained in the mould, the artisan had open in front of him a book with a picture of the original sixteenth-century pot in order to correctly add the finishing touches to it. He was therefore contributing more work that the simple reproduction of pieces.

Another instance where the name on the pot is not the name of the maker is forgery. Because the name of the maker can make an important difference to the value of a pot, forgers simply reproduce the seals of artisans and stamp them on their own work. Forgeries have always been a problem in the Yixing 'art world', and appeared soon after *zisha* pots became luxury items in the sixteenth century (Bartholomew 1990). I have mentioned that during the 1930s, as a result of the boom in demand for antique teapots, shops in Shanghai started employing talented artisans from the Dingshu area to reproduce, or sometimes 'invent', antique teapots. Officially, the name stamped on the base of the pot was to be the maker's, but mostly, the market became swamped with objects of very high quality, claiming to be the work of Ming and Qing masters.

Today as much as in the past, insuring authenticity can be difficult, and collectors, as well as the artisans themselves, have developed certain techniques to guarantee as much as possible that their customers are getting an authentic work. As mentioned, one strategy used by artisans is to insist that customers visit them personally; this might not put a stop to forgeries, but at least the artisan can claim that any work not purchased directly is likely to be a forgery. In cases where an item was not purchased directly, many collectors will go to Dingshu to determine its genuineness, and ideally they will ask the presumed maker directly; if the artisan had died in the meantime, then the latter's student might help.

My informant Pan Chiping has, as I noted, been conducting historical research, and has become a reference on determining the genuineness of antique and contemporary teapots. Many collectors visit him to show him an item in their collection, and even museums consult him. Pan has several ways of identifying original and fake pieces. For instance, some historical abnormalities might suggest that the pot is a fake. One pot he was given to be evaluated, for instance, supposedly from the Ming dynasty, had an inscription containing the name of the artisan alongside the name of the emperor at the time (to indicate a date). Pan pointed out that the artisan would have got into trouble by placing his own name at the same level as the emperor's, and concluded that the pot was a fake. But in other instances, Pan claims he is able to determine the genuineness of a pot, saying 'I look at the style' (*wo kan fengge*). As an example, he showed me several pictures from a Qing dynasty artisan's work, and pointed out certain similarities. For instance, the end of the spout would be rounded on some, squarer on others. Now, this might indicate that the style of the artist changed, and indeed the works spanned more than two decades. But sometimes, there are certain abnormalities that appear, where details from two contemporary pots differ significantly, and the style of one pot does not correspond to that of a body of work. In another pair of objects, the base of a pot is nearly identical in two examples, indicating that the two had been made using the same tool – and therefore most probably by the same person. Pan was pinpointing certain details, but overall *fengge* is not something that was limited to details. I think this was most forcefully put forward by Zhao Jianghua, who claimed to be able to recognize with the shortest glimpse whether a pot had been made by his late master or not.

Alfred Gell argued that one should consider art objects to have agency. An object's agency results from an act of abduction, that is 'a tentative and hazardous tracing of a system of signification rules which allow the sign to acquire its meaning' (Eco, quoted in Gell 1998:14). 'Hazardous' is the key term here. Gell is not so much concerned about whether or not things might be extensions of persons and their agency but how things are perceived by others as actually being another agent. In the case of handwriting, the elements that are inferred are the inner dispositions of the person behind the pen or calligraphy brush: Yen Yuehping's informants told her that they often formed an initial impression of a person from their handwriting – an impression of the person's morality, character, and even looks (Yen 2005:64–72). Determining the genuineness of a pot involves proving or disproving the claim the name stamped on a pot makes as to the origin of the object. Pan and Zhao inferred authenticity from signs they perceived as significant in objects. *Fengge* is the term of this group of significant signs, and to claim that this *fengge* testifies to the genuineness of a pot is an act of abduction. This is of course a problem in many arts, but in Dingshu, I was surprised at the way the authority to determine such authenticity relies on *fengge*, and the idea that some pots can be identified beyond doubt from the signs that the 'body-person' imparts on the surface of the clay. Artisans who had prolonged (visual) contact with the presumed author of the work being evaluated are expected to be able to determine authenticity simply by looking at the pot. Where is the guarantee that the resulting opinion is correct? Regardless, an 'authority,' by putting forward an opinion, is in effect attributing an identity to an object – and probably at times forcing one onto it. I addressed one such case in Chapter 3, where a master reportedly passed off one of his own creation as an antique Qing dynasty pot in an authoritative book on *zisha* pottery he co-authored. How many other such successful forgeries might be found in museums and private collections? And in how many other cases is the identity attributed to the pot not that of the artisan?

Conclusion

I talked in the previous chapter of the craft of making *zisha* ware being engaged with the body, which in turn leads to the personal development of skills and to their very inclusion in the body's make up, in the idea of *gongfu*. It is a short step to take to suggest that a link of identity is created between the resulting work and the artisan, and indeed *fengge* is understood as such a link. The resulting style of *zisha* craft is understood as a by-product of techniques and skills perfectly mastered over the years: they testify to the actions of the artisan in the course of making the object. This link between artisan and creation is subsequently reinforced in the act of stamping a name on a pot. Yet this explanation is a short cut: these 'signs' that seem to testify to actions, and to the link between the craft object and an artisan, might be too easily made. The fact that there is a link of identity between artisan and creation is an abduction of features that are present in the pot. Yet as artisans say themselves, it is up to others to determine what this link is; it is up to others to recognize it. Since other people's gaze has a constitutive effect on such a link, I see this link not as some emanation of the person or character but as socially constructed.

A problem arises regarding agency. If we think of the pot as a secondary agent – to use Gell's (1998:20–1) terms – who is the primary agent? Other people become the agents that create the link between artisan and creation; one might then say that there is more than one 'artist' (to use Gell's terminology), or originator, in the making, not so much of the object but of an individualized (named) object. The artisan needs others to talk about them and their *fengge* in order to exist in the *zisha* art world. As Gell (ibid.:33–4) notes, 'patients' (spectators) can also be primary agents in the art nexus. Webb Keane (2001:74), talking of Sumba, in Indonesia, has suggested that the capacity of objects to serve as representatives of persons requires constant effort to sustain it. I think the word 'effort' is useful in thinking about my own case study. Techniques in *zisha* pottery are ultimately aimed at stamping the self onto a clay object. There is something magic-like in this process, and indeed this capacity transfers agency to a pot, an agency that will be able to present to others a guarantee of authenticity. But there is also a fear that the object might lose its origin, its link with the artisan; therefore, this link must be sustained with a number of strategies. *Fengge*, individuality, is something that needs to be extracted through hard work; the signs of the person must be forced out, in the endless repetition of 'correct' form. And the identity will further be forced onto the object, with the seal (which is, appropriately, hammered onto the clay) and attributions of authenticity. In Dingshu, work should not be alienated nor dissociated from artisans lest it lose all value. Instead, identity must be forced onto the object, the object must be made inalienable. It is authorities, and I would add charismatic figures, who take on the role of enforcing the relation between a person and an object. It is these charismatic figures, the masters, that are the subject of the next chapter.

CHAPTER 7

Mastery and singularization

So far in this book, I have followed two broad lines of enquiry. I have firstly been interested in the making of craft objects with skill, and have addressed the processes of learning the techniques and designs of *zisha* craft. In parallel I have been concerned with the national and local politics of craft production, how crafts were caught up in the social and economic transformations enforced by the communist government, as well as the internal politics of the community of practice of Dingshu ceramics. At several moments in this book, these two lines of enquiry have necessarily become intertwined. I have, for instance, talked about the advantages of having been a worker in the state-controlled Zisha Factory for gaining fame and reputation in the *zisha* art world (Chapter 3), the role of the Chinese state and its national intangible heritage lists in the *zisha* art world (Chapter 3), the social capital of artisans, acquired in a cooperative factory, that enables them to access certain exclusive forms of knowledge (Chapter 4), and the continued use of traditional tools, which implies both continued commitment to certain craft techniques and the possibilities for creating exclusivity in the craft's regime of value (Chapter 5).

This chapter brings together perhaps more forcefully the skilful and the political in a discussion of mastery. A master is, by definition, both a person who has attained the highest levels of expertise and an influential figure regarded as a model artisan to be emulated not only technically but also, as I will show, morally. Anthropological monographs on craft occasionally devote a final substantive chapter to the topic of mastery, or even to individual masters. The last chapter of Eugene Cooper's monograph on Hong Kong woodworkers introduces us to the master Cheung Kungngai, a highly skilled artisan trained in pre-war China, who had managed to continue his art in the face of competition from mechanized production, but who faced the prospect of returning at any moment to wage labour (Cooper 1980:125–35). A discussion about this master enables Cooper to reflect on the changing features of the woodworking industry. The master is the epitome of tradition, an individual who has attained a high level of skill, yet whose skills are rendered almost worthless with the arrival of mechanization. Trevor Marchand, in his monograph on minaret building in Yemen, concludes with a discussion of mastery (Marchand 2001:184–239). Chapter by chapter, the book follows the erection of a minaret from its foundation to the construction of the dome. Each stage of the construction process corresponds metaphorically with a stage in the career of a builder, from worker to apprentice to master. The master is the individual who has not only developed technical knowledge, but who has also acquired understanding of the principles, technical and aesthetics, of the minaret building tradition.

But what exactly is a master, and how does one become one? Despite attention to masters as persons who have attained high levels of expertise, there remains a lack of critical attention to the category of master in the anthropological literature, though there are some notable exceptions (Lancy 1980; Marchand 2001:184–239, 2009:167–88). Being a master is more often than not taken to be coextensive with 'expertise'. In their classic definition of expertise, Dreyfus and Dreyfus (1986) define the expert as a person who, through long practice, has gained a level of knowledge whereby new situations are addressed through intuition rather than rote reproduction of past situations. A master is an expert, certainly, but a master is not only an expert but also a person whose expertise is recognized by fellow artisans, who come to emulate their ways of making, designs and sayings. Ultimately, a master is a significant node in the making of a tradition, a person who comes to redefine elements of a tradition.

I have suggested that over the past few decades, *zisha* pottery has been reinvented as traditional. By this I mean that, despite the dynamics of innovation and change that exist in any craft, artisans have strived to create an illusion of stability, referring to a 'tradition' (*chuantong*). The masters of *zisha* pottery, I want to suggest in this chapter, are instrumental in creating stability through change. I conceptualize the role they take as that of 'editor' of the craft, individuals who are innovative and creative not in producing new designs and techniques, but in creating apparently unchanging works. In a nutshell, my argument here is that a master needs to be recognized as a social phenomenon rather than simply as a highly skilled individual.

Lack of attention in the anthropological literature to the social significance of masters, beyond their capacity to pass on their knowledge to the next generation, might be due to a problem of methods. At least since Michael Coy's seminal edited volume on apprenticeship in theory and as method (Coy 1989b), ethnographers have enthusiastically adopted the position of novice in the field, learning about technical processes and acquiring skills as a privileged position through which to understand the people that are the subject of their studies. Together with the turn to embodiment in anthropological theory, this has also provided anthropologists with understandings of the tacit and experiential aspects of the lives of informants (Jackson 1983). Interest in apprenticeship as a research method has naturally shifted the focus towards the learning experiences of apprentices. Scholars such as Erin O'Connor (2006) and Loic Wacquant (2006) provide rich descriptions of the sensorial and emotive dimensions of their own experiences of becoming glass-blowers and boxers respectively.

But what about accessing the experiences of master artisans? Keller and Keller (1996) provide what might be the best account of the experiences of experts: Charles Keller's expertise in blacksmithing offers us a rare glimpse into the making of things with skill, an example where a professional anthropologist talks about a craft in which he has attained high levels of expertise. Here again, although very insightful, Keller and Keller's discussion is about expertise, not mastery. Marchand (2003) even argues that the aim of the anthropologist of crafts should not be to discuss the master's knowledge since it is impossible to have access to such knowledge, which is not semantic and thus not expressible in words. Rather, for Marchand, we should be concerned with understanding the nature of knowledge and the human mind, what it means to have knowledge in the first place. To make a perhaps clumsy division between expertise and mastery, when we talk about expertise we are talking about skill, about tacit knowledge. Mastery, I want to argue in this chapter, is about the social, it is about recognition of the expertise, as well as the related knowledge, including esoteric knowledge, of certain individuals called 'masters' by other individuals – other artisans, connoisseurs or, more generally, members of an 'art world'. This chapter thus addresses the person of the master in social terms. It considers expertise not simply in making things, but in changing the way these things are perceived, made and talked about by others. In sum, I am concerned about the specific kind of agency achieved and deployed by masters. Since I cannot rely on personal experience of mastery, I have to rely on the more traditional data available to anthropologists, namely words, especially in the form of narratives. Putting aside concerns about phenomenologically accurate descriptions demanded by the recent theoretical turn, I take literally the experiences of mastery as narrated by close associates of a master. These narratives offer as a kind of magical expression of what making things with highly developed skills might be. I was told about a master who could talk to clay, who could make his pots talk and who made objects with the precision of a machine. These narratives serve as my point of departure for understanding the significance and influence of a master of *zisha* pottery, and to think about what ultimately can never be explained: creativity and genius.

Who becomes a master? What kind of path through life leads to being revered as a master of one's craft? Of course, there are as many answers to those questions as there are craft masters. Masters need to be unique in order to be recognized as such. Masters are 'singularities', they are individuals who are unique and cannot easily be categorized, and for that reason open up avenues for creation and innovation that were not previously thought possible. I am certainly romanticizing the master here, but once the hagiographic narratives have been told, I will turn to a more critical analysis of the role of the master in a community of practice and an art world. To illustrate my arguments, I take the life of one particular master, Gu Jingzhou, usually spoken of as the greatest master of *zisha* pottery of the twentieth century. Necessarily, his story is unique, but it reveals much about the role a single individual can play in shaping the trajectory of a craft.

The masters of *zisha* pottery

For connoisseurs of *zisha* pottery, and for the artisans of Dingshu, Gu Jingzhou is the exemplar of the *zisha* artisan. I have mentioned several times the figure of Gu throughout this book. Gu was, in the minds of virtually all those I talked to, simply the greatest master of the twentieth century. Listening to people talking about him, I often longed to be able to turn back the clock to a time before his death in 1996, seven years before my fieldwork, in order to meet him. He was described not only as a highly skilled artisan, but also as a wise and highly knowledgeable individual who could talk at length about the craft, as well as Chinese culture and history. I came to know well many of his former students, and through their stories the figure of the master took on life and meaning. During fieldwork, I soon realized it was important for me to attempt to reconstruct the life of this master from the accounts of his former students. More than simply reconstructing a life story, I became increasingly interested in the way these narratives were weaved into the practice of everyday craft production in Dingshu, how his work and words came to be reproduced by contemporary artisans to justify their choices and outlook on techniques, design and tradition. I choose the case of Gu as one I know well, and also, more importantly, because he was the person most talked about in Dingshu. In talking about the master, artisans strived to establish a connection with him – having been a friend, a student – and in so doing they infused their own work with the aura of the master. By looking at Gu's career, his sayings and his innovations, one gets not only a better understanding of *zisha* craft, but how the craft might be renewed, and more specifically, how an individual might redefine a 'tradition'. In that sense, my concern here is to address the dialogic between master and tradition, the potential role an individual can have on a whole corpus of knowledge and skills.

The word 'master' (*dashi*) is commonly used in Dingshu and carries certain specific meanings, but it is also an official title that can be obtained from different government institutions, either at provincial or state level.[1] These are the 'official' masters of *zisha*, but as might be expected, such titles are often given for political reasons rather than as recognition of excellence. Today's official masters are those who were among the first to join the Zisha Factory, between 1954 and 1956; most came to occupy high managerial positions in the factory. Talk of buying titles, and nepotism in obtaining them, are a not uncommon part of local gossip. One artisan even claimed that one of the current titled masters was quite incapable of making a pot. But 'master' is not only a title but a way of imaging certain individuals thought to be more special than others. When I asked my informants what distinguishes the true master from the common artisan, a recurrent answer I received was that 'the master is someone who can make, speak and write' about *zisha* craft: *huizuo, huishuo, huixie.* I will briefly consider these three capacities in turn. The first is evident. It denotes high technical skill, though I will have more to say about the ability to 'make' further on. 'Can speak' might be better rendered as 'can explain'. I was told that, although experts can recognize whether a pot is good or not ('good looking', *hao kan*), a master can do more than that: they can explain in what way a pot is good or bad. This might be in terms of the properties of the design, of lines, of the balance of forms and so on. One artisan explained to me that a master is one who knows the theory (*lilun*), in other words a profound understanding of the craft. This means being able to identify and describe the fundamental principles that should govern the craft. In a nutshell, masters are able to argue about these principles, with reference to specific objects and technical issues as well as aesthetic canons.

It must be recalled that apprentices in the old factory, and still today, learn about the theory of *zisha* craft. Rendering into words the aesthetic and technical principles of the craft was a major innovation of the Zisha Factory, together with teaching in classes of twenty or so students, one which was perhaps even required by this new configuration of teaching. Students were taught about balance in terms of ratio, of the lines of a pot as forming circles or a parabola, or the aesthetic principles behind the matching of a spout with

---

1 There are a number of titles artisans receive before that of master, with an optional title of 'famous person' (*mingren*)!

a pot's body. However, in the words of my informants, masters are individuals who not only have that knowledge but also true understanding,[2] which in turn enables them to formulate this understanding in speech, to make it explicit.

The third capacity of the master, writing, might also come as a surprise, as we are used to a division of labour in Western arts between the person who produces art (the artist) and the one who writes about art (the critic). This refers to the ability to write about the culture of the craft, in the form of essays published in specialized reviews and destined for a public of experts. Writing articles in specialized periodicals, on tea or ceramics connoisseurship, is so important that it has become a formal requirement for promotion to the title of master. Publishing implies becoming a well known figure in the world of *zisha* connoisseurs, and therefore encompassing the dual roles of artisan and connoisseur (see Chapter 3). It also implies, importantly, that one's ideas can potentially become influential. In fact, in support of this last point, some of today's masters are those who had taken managerial positions in the Zisha Factory, and as a result almost stopped producing; the title of master is largely defined in terms of influence in such cases and is dependent on the politics of the craft community. Contemporary artisans who have received their title at provincial rather than national level are those who have lost the favour of the more powerful individuals of the community.

Artisans pen a variety of texts, about the history, aesthetics and archaeology of the craft. Some articles published in the journal of the Yixing Ceramics Industry Association simply give an account of the life achievements of a fellow artisan. Gu Jingzhou wrote mainly about the aesthetics of the craft, but he was also an authority on writing about history. One of the main collections of articles on *zisha*, edited by the head of the Yixing Ceramics Industry Association located in Dingshu, regroups several articles by Gu Jingzhou, as well as some by other authorities on the craft, whether artisans or not (see Shi 1991). The articles address the craft in general, highlighting its specificities, or talk of certain individual artisans, their life and style. One of the essays is penned by Gu Jingzhou, who takes it upon himself to define the crucial aesthetic principles of *zisha* (Gu 1991). These are form, spirit and energy (*xing, shen, qi*), terms that are common to texts on the connoisseurship of calligraphy (see Billeter 1989; Yen 2005). The article is particularly significant since it weaves together the particularities of *zisha* craft and the language of culture related to one of the highest forms of Chinese arts, calligraphy, and thus roots the craft ever so strongly in high Chinese culture. In this article, the

---

2 Marchand (2001) defines mastery in a similar way when talking of Yemeni minaret builders. His case is however rather different, since apprentices are trained without being given any form of explanation.

master also brings us to see the craft in a certain way, not simply as a collection of clay objects made with skill, but a tradition intrinsic to Chinese civilization. I now turn to a closer analysis of the process of becoming a master in Dingshu. In this respect, the artistic, personal and political path of master Gu Jingzhou offers both an illuminating and a unique case study.

Gu Jingzhou's atypical career

Gu Jingzhou started his career as a potter in the 1930s, and he reached the height of his creative career in the 1990s. His career thus spans some of the most tumultuous decades of recent Chinese history. Gu was born in 1915 in a village in the vicinity of Dingshu. His father had a successful pottery business employing artisans to produce *zisha* pottery. Gu started to learn pottery when he was 18 – a relatively late age, for at the time, and during much of the period of the Zisha Factory, apprentices would start at the age of 14 or 15. However, Gu's father had wanted him to have an education, and he had been taught by a private tutor hired by his father. Gu claimed that he learnt how to make pottery from his paternal grandmother. Perhaps more likely, however, and according to at least one of his former students, Gu's master during his apprenticeship was one of the artisans employed by his father. In any case, he was soon found to be particularly talented – my informant explained this by suggesting that he had watched artisans working for his father every day and from an early age, and given that his father's business made high quality pots, Gu knew much before he even started his apprenticeship. He was talented enough to find work in Shanghai very soon after the end of his three-year apprenticeship; at the time, in the 1930s, the best known shops that produced *zisha* ware were located there, and they only employed the best ceramics artisans from Dingshu.

I have described in Chapter 2 how artisans from Dingshu were employed by these Shanghai shops to produce forgeries of antiques, and it is in this context that Gu began his career. For Gu, this was an invaluable opportunity to improve his skills and learn about classic designs. Artisans in Dingshu rarely had the opportunity of using antiques as models, but in the context of the Shanghai shops, artisans had access to reproductions of classic designs, and in some cases to actual antiques in private collections. The forgeries had to be of high quality, and the emphasis was on quality rather than quantity: artisans would sometimes take months to work on and perfect a single piece that could be passed off as genuine. Forgers also had a fair degree of freedom in creating pots. For a forgery to be credible, artisans could not simply reproduce an antique pot in a known private collection; rather, artisans were to create new, unique designs, and rely on their imagination to know what an unknown and unique antique might look like. Artisans had to know about the aesthetic and technical characteristics of ancient pots, but also demonstrate creativity in producing variety. The possibility of working on quality rather than quantity was without doubt a great asset in the development of Gu's career. In Shanghai, Gu was also part of a circle of scholars, who discussed art theory and Chinese culture. These formative opportunities were rarely available to artisans in Dingshu.

During and after the Sino–Japanese War (1937–1945), production of fine *zisha* pottery came to a standstill, and Gu, along with many artisans, was forced to abandon his craft. Together with his first apprentice, he took up wage labour in the factories of Shanghai. Gu was later one of the first to join the Zisha Workshop. In the first years of the Zisha Workshop, Gu developed the core of his teaching and technical principles, most probably through discussions with a scholar from Beijing's central art school, Gao Zhuang, who had travelled to Dingshu to commission pots based on his own designs, as well as introduce such innovations as the plaster mould to the cooperative factory. Gu became close to Gao, and much of the content of Gu's teaching curriculum, known as his 'eighteen lessons', which encompassed technical and aesthetic principles of the craft, was derived from the ideas of Gao. As one of my informants stressed, not all of Gu's ideas were his own, but Gu Jingzhou was instrumental in disseminating among common workers ideas that until then were only circulating in scholarly circles and among connoisseurs. These principles were innovative and modern in scope, and explained the aesthetics of *zisha* pottery in rigorous geometric terms, including references to parabolic lines and ratios between the different elements of a pot.

At the Zisha Factory, Gu was not only a talented worker but also a designer and innovator. At the head of the research team, and during the difficult times of the Cultural Revolution, when export links were cut, Gu and other workers sought to develop alternative products to keep the factory going. Gu is said to have developed water-filtering vessels using *zisha* clay. I am told that the problem-solving approach he took then resembles the approach he took to potting – I will introduce an example of this later.

Gu remained a worker at the factory after the economic reforms. As an incentive for Gu to remain, the factory managers agreed to take a much lower commission on his products. In the 1980s, he moved from the open spaces of the factory to one of the private workshops that the factory set up for masters, and he worked there with one, and later a second student. After he retired, Gu took up quite fervently the practice of calligraphy, and read extensively about Chinese history and culture. In China, it is quite common for pensioners to take up such practices. In Dingshu, after they retire, masters cover their work table with a cloth, transforming it into a scholar's desk on which to practise calligraphy and read. What I find interesting is that master Gu's students link the high level of culture he achieved to his capacity to talk about his art with an unprecedented expertise, and the few works he produced during that time are said to far surpass all his previous work.

It is worth taking pause here to think about the relationship between the practice of calligraphy and the master's practice. The practice of calligraphy holds a central place in personal self-cultivation in China. I have mentioned that the anthropologist Yen Yuehping (2005) identifies the acquiring of *wen*, 'culture', as central to the notion of Chinese personhood, and the practice of writing and calligraphy is one important way in which culture is acquired. What I think Gu Jingzhou's students were implying when they stressed that he had reached the height of his genius only after he had started practising calligraphy, was that his self-cultivation as a person, including in moral terms, became apparent in his work with clay. This idea is also related to the notion of personal style, *fengge*, that the pots of only proficient artisans are said to have, as discussed in the previous chapter: the moral and cultured properties of persons transmitted to the objects they create.

Gu remained in his factory workshop until the last years of his life, relying on one of the factory's doctors (Zhao Jianghua, who would later become his apprentice) to look after his declining health. Gu did not have a family, and relations with his students seemed to be substitutes for family relations. His students would help him take a bath every evening when he no longer had the energy to do so, as would any elderly person's sons or daughters. Gu Jingzhou died in 1996, six years before the privatization of the factory that he had helped create.

Gu Jingzhou was the reference point for so many aspects of the craft, not only on the correct way of reproducing certain designs, but also, more generally, on how one is to understand, and talk about, *zisha* culture. He was also a reference on *zisha* history, and collaborated with archaeologists on interpreting pottery shards found in the region, and wrote many influential essays. No one in Dingshu would deny that he was the greatest master of the twentieth century. This was also recognized at national level: in 1988, and shortly after he was convinced by the Zisha Factory's managers to join the Communist Party (which he is reported to have had disdain for), Gu was endowed with the prestigious title of National Master of Arts and Crafts by the central government, becoming the first and to date only holder of the title among *zisha* masters. The authority that Gu attained as a master is to be explained with reference to many factors: the politics of the factory and the instrumental role he played in establishing Zisha Workshop, his personal achievements as a potter, and himself as a self-cultivated person. He also had tremendous influence as a teacher, and taught many among the newer generations of artisans from the early years of the cooperative until a few years before the privatization of the factory. He introduced innovations in the teaching curriculum: he had taught the craft – as well as a way of talking about the craft, a way of looking at it – to hundreds of students over the years of the factory; his way of understanding the craft had in effect become institutionalized. And as a person, over the last two decades of his life, he developed himself as a moral person, practising calligraphy and learning about Chinese history. Tseng Yu-Hui (1998), a connoisseur of fine Chinese teas, goes as far as claiming that Gu was simply the last of the *zisha* artisans, the last to be versed in the techniques of the Ming dynasty, and as a result the craft died with him in 1996. I need not spend time arguing against such a short-sighted view – if a craft must die with the last surviving master, then *zisha* must have died many times during its history. The statement does, however, give an idea of how connoisseurs perceive Gu's significance for the craft as a whole.

Narrating the master

Zhao Jianghua, who was a student of Gu Jingzhou, once told me that the pots his master made were so perfect that they looked as if they had been made with a machine. Zhao was here likening the high level of skill of the master to a trick he and his fellow artisans often resorted to. In order to get the mouth of the pot perfectly round (one of the basic requirements of a good *zisha* pot is that the lid fit perfectly, and turn on itself without a snag), after the first firing, the mouth is filed with a machine, a lathe, and then the work is patched up with a little clay before a second firing. Gu, however, never needed to resort to this trick, because his pots reportedly came out perfect after the first firing, as if he had used the same machine. The image of the master as machine was one of the many ways in which Zhao tried to convey to me the genius and technical excellence of his master.

Many narratives, such as this one, were recounted to me to convey the superior qualities of Gu and explain his status as the greatest master of *zisha* pottery. The narratives at times take on quasi-mystical or magical dimensions. The capacity of making a pot perfect is magical to even expert artisans who know of the difficulties of the craft, and who have to rely on extra steps in order to produce a pot of high enough quality for sale. The image of Gu as a machine is particularly striking, suggesting an artisan with almost superhuman skills. For the sociologist Richard Sennett (2008), machines constituted a threat to artisans of the eighteenth century as they represented something that could never be attained by human strength and skill. The machine was supposed to supplant the indeterminate manual skills of the human body, yet in the image conveyed by Zhao's narratives, the master becomes that machine. The image of the master as a machine might appear surprising to a Euro-American public, maybe even morally wrong. Indeed, the image radically contradicts some of the guiding principles of the Arts and Crafts movement that originated in the nineteenth century in Europe, ideas about craft that have been carried into the twenty-first century; Euro-Americans tend to assume fundamental technical differences between craft and industry. But this difference, as Pye (1968) notes, is a moral one, since he argues there are no substantial differences between craft and industrial production; for Pye, even the concept 'handmade' breaks down under scrutiny. The comparison of a master's skills with a machine did indeed strike me, yet at the time, well into my fieldwork, I had grown accustomed to artisans talking about the high level of craftsmanship of their pots, and the way that the quite unique techniques used were geared to achieve the high level of precision that customers required. However, the main notion that Zhao was trying to convey with the image of the master as a machine is not, I would argue, the mechanistic skills of the master but the notion that the master was able to render his intentions in material form effortlessly. This idea prevails in the narrative about Gu Jingzhou that I heard from his former students, and I want to spend some time unpacking these narratives and ideas.

I have talked about the relationship the artisan has with the objects they create in terms of *fengge*, and the attribution of *fengge* by connoisseurs. What about the master? I was told that a master can do more than simply make a beautiful pot, one that also has *fengge*. Zhao told me that there is a saying in the local dialect to the effect that some pots can 'talk': *hu neng shuo*. However, it is only the greatest of masters who are able to make a pot that 'talks' in such a way. Gu Jingzhou was such a master; towards the end of his life, he would tell his students that when he was making a pot it would listen to him speak (*hu ting wode hua*, 'the pot listens to my words'). The idea is that if the master can 'speak' correctly to the pot, the pot 'listens', and at a later stage will be able to 'speak' when someone comes across it. What does it 'speak' about? Of 'history, culture, and self-cultivation' (*lishi, wenhua, xiuyang*), according to Zhao. The ability to 'speak' can be understood as the ability of the object to embody a set of values – in many ways, it becomes a miniature, in Lévi-Strauss's meaning of the term (Lévi-Strauss 1966:22–5), of some core values of Chinese culture (see Chapter 3). Gu talked of giving life to a pot (*renxinghua*, literally 'making human'). The common artisan, in contrast, might try to 'speak' to the pot, but the pot simply cannot hear. To illustrate this point, Zhao told me an anecdote. Once, a scholar came across one of Gu's pots, and after observing it, said 'this pot can talk' (*zhe ba hu neng shuo hua*). That it was a scholar who 'heard' such things is important. Indeed, it is not given to ordinary people to 'hear' a teapot! A commoner might recognize and remark on the beauty of the object, but only a few can really understand it. In the words of Zhao, this scholar could properly 'see' (*neng kan qilai*).

There is an interesting blurring of registers in the 'magical' narratives told about the master. The scholar 'sees', and therefore 'hears'? How should one understand this confusion of the senses? The pot might be able to 'speak', but not in the sense of telling a story; there is no message to decode, as in a structuralist interpretation of visual form, and it is not meaning that one derives from looking. Rather, I see this 'speech' as the capacity to reveal, in the sense of the miniature, the fundamental properties of the craft. If this is explained in visual terms, why then the metaphor of speech? One might be tempted to refer to a notion that has gained popularity in the anthropology of art, Alfred Gell's idea of the 'abduction of agency', whereby a work of art carries on the agency of its maker in the latter's absence (Gell 1998:13–16). The agency of the master, in this case in the form of speech, is transferred to a 'secondary agent', the pot, that metaphorically at least is said to be capable of speech. This speech can in turn be transmitted to (or affect) another person, in this example the scholar. The notion that the pot can talk of 'self-cultivation' is intriguing. I believe it is to be taken as that quality that the master has developed, in his perfect comprehension of the principles of the craft and engagement with Chinese culture, in particular through the practice of calligraphy and reading about history. In that sense, and continuing with the analysis inspired by Gell, there is another abduction of agency, whereby the quality of the person is transferred to the object: a kind of extended person, it contains the qualities of the primary agent that are transmitted by contagion. This quality that is transmitted to the pot is different to that which I described for personal style, *fengge*. Yen (2005) talks of the morality of a person that transpires in the written characters they write. The case we are concerned with follows the same principle. I have mentioned that Gu's students claim that his best work was made in his last years of life, after he had started this self-cultivation, studying and practising calligraphy – the link seems clear between the self-cultivation of the master and the excellence of their work.

Gu said he talked to the pot. In this image, of the master talking to clay taking shape, there is also another suggestion of the quasi-magical skills of the master. What is implied in this image is a kind of sublimation of technical mastery. The image Gu used was one where thought and speech alone could produce a pot, where the skills and techniques he had mastered over his career were somehow surpassed and no longer needed. Another image Gu used to describe his own work might make his ideas clearer. Gu used to say that when he made a pot, the clay was like a child: he would tell it what to do, 'go left, go right', and the clay would obey his every command. The relationship between master and clay here is then one of authority, the authority of a father over a child, and 'speaking' to the pot implies ordering it about. In the image Gu used, it is these orders, this speech, that shapes the clay, rather than the hands or tools. Gu was, however, not implying that anything he made came out perfect each time. Zhao told me how once Gu remade the spout of one particular pot eighteen times to get it exactly the way he wanted. But according to Gu's sayings, there seemed to be a perfect coordination between thought and action, so that an idea could be effortlessly translated into clay form.

One might recall here Yen's notion of the body-person in calligraphy (Yen 2005:75–80). I would suggest that in the present case, mastery involves the perfect unity and coordination of the body-person. The major contrast with Yen's study, however, is that in the case of *zisha*, the expression of the body-person does not reveal inner dispositions – Yen mentions moral standing and character in particular – but rather perfect adherence to the principles of the craft. When Gu Jingzhou made a pot, and at least according to the narratives artisans tell about the master, the object became the ideal materialization of that particular shape of pot. In the previous chapter, I mentioned how Zhao could immediately recognize a pot made by his master by looking at its *fengge*. In one instance he told me that when a collector showed him and other students of Gu a pot in order to decide whether it was indeed made by the master, they looked at, among other details, the interior of the spout. Seeing the perfection of craftsmanship, all agreed the pot was indeed authentic. What is intriguing here is that Gu's *fengge* should be recognized in the perfection of craftsmanship. These are not signs of the person, yet it was a 'signature' – in Wollheim's (1979) sense of the term, as signs that enable the identification of a work of art – that guaranteed authenticity.

I have suggested that the comparison of the master's skills with a machine are to be understood in terms of perfectly rendered intentions rather than mechanistic skills. In turn, the master's intentions are the rendering of the perfectly understood principles of the craft, and in the hands of the master, the finished product becomes the ideal materialization of these intentions. One might say that the master is able to transcend their own personal style, or *fengge*; what is left is the perfect embodiment of what are perceived as the 'traditional' core principles of the craft.

Perfecting traditional designs

Gu Jingzhou is not known for having created original designs (though he did collaborate with scholars and artists on new forms). Instead, he was known for work on what are described as 'traditional' (*chuantong*) designs, usually geometric and simple in form. The *fanggu* pot ('in the shape of a drum') is one example Gu is famous for. It is very interesting to look at different versions of this shape as made by the master, and to notice a very clear evolution in it (his students say they can recognize the period in which an example was made by its shape). Some elements became more significant for Gu in later years. The 'shoulder' of the pot, for instance, becomes more pronounced as the years pass, and the space between the shoulder and the mouth of the pot becomes perfectly flat in the *fanggu* pots he produced in the last decade of his life. Such elements make his pots recognizable. Interestingly, different students of Gu will make a *fanggu* according to how Gu made it at the time they were studying with him. So the *fanggu* of the first of Gu's students is different from that of later students. Numerous photographs of versions of the *fanggu* made by Gu have been published, so that the association between master and reproductions has been reiterated and strengthened with every publication.

The changes brought by Gu to such shapes as the *fanggu* were evolutions in the strict sense of the word: Gu said that he was 'improving' (*tigao*) the design, so that it reached 'perfection' (*wanmei*). He claimed that he had achieved the perfect rendition of the types of pot he had worked on, and had the assurance to claim that once a form was perfected, no one else could find a better way of making that particular pot. Gu would say of the designs he put down on paper, 'the designs I draw cannot be changed' (*tuzi bu neng gai*), meaning that if anyone finds a problem with the design, it is their own problem, not a problem of the designs: he would say, 'it is your problem, not mine' (*shi ni de maobing, bu shi wode*). His successors are faced with the choice

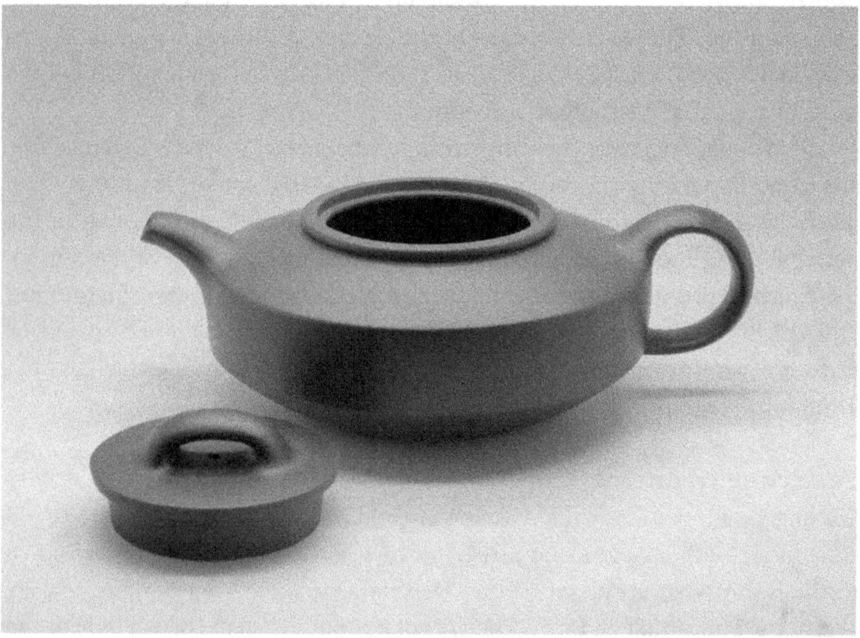

Figure 17 Gu Jingzhou's 'improved' design of a classic teapot (made by Zhao Jianghua). The open lid reveals how a 'neck' or rim was added to the body of the pot. In the classic design, the mouth of the pot was rimless, and the lid, made of a single disc of clay, rested directly over the mouth of the pot. In the new design, when the lid is in place the pot retains the appearance of the original design.

of either reproducing the pots as he instructed or looking for other classic forms to 'perfect'.

Zhao talked of Gu having achieved no less than what he termed *zhenli*, which can be translated as 'truth'. How does this perfecting of shapes operate? I learnt of an example of a design perfected by Gu, which illustrates his idea of improving a design. Zhao told me that once, when he was learning under Gu, he thought of reproducing a classic design from a picture of a pot he had found in a book. Gu encouraged him to do so, but suggested he himself redraw it because there were certain 'mistakes' (*cuowu*) in the original. He introduced one particular change, that although minor – in fact it did not even change the outer aspect of the pot – made the design 'perfect' (*wanmei*). It consisted in changing the way the lid was made. The new design is shown in Figure 17. In the classic version, the lid is flat, and when put on the pot, the edges rest on the edges of the mouth. Gu Jingzhou changed this. The mouth of the modified pot has a ridge, and when the lid is fitted, its surface is flush with the top of that ridge. The result is rather surprising: in effect, nothing has changed, in that the profile of the pot is exactly the same (at least when the lid is fitted). Why then was the modification needed? It must be recalled that a pot is described in terms of features of the human body. Artisans for instance talk of the neck, shoulder and belly. For Gu, the original design did not respect these principles: something was lacking, or rather the parts did not appear clearly enough. The problem concerned the area where the rim of the lid touches the body of the pot: made in the classic way, the line is 'ugly', and not in the control of the artisan, because it is simply the line created by the rim as it rests on the mouth of the pot. In the revised design, the artisan can act on that part, because the body of the pot and the rim can be made in one continuous line. The perfect pot must underline the difference between the body and the rim, with a thin but clearly visible line separating the two. In more geometric terms, the two convex lines of the body and of the rim should be separated by a barely visible concave line. The change is not only aesthetic, it is also structural. The mouth of a pot has a greater risk of shifting shape during making if it is not 'supported' by a rim added on top. And when used, there is less risk of breakage.

An intriguing parallel is to be found in the anthropological literature. Scoditti (1990), writing on the Kitawa islanders of Papua New Guinea, a group famed for the carving of the prow-boards of some of the canoes that take part in the Kula ring, tells us about a similar reasoning on the part of a master carver. Scoditti tells us that there are two sorts of artists on Kitawa, simple carvers and master carvers. This distinction does not refer to the degree of mastery over the carving process, but to the esoteric knowledge that the master carver has obtained from his own master in the latter's old age. This knowledge concerns spells and the relationship between the elements of a design. The master carver is, in that way, at greater liberty to introduce innovations than other carvers. Master carvers are at liberty to either modify the interpretation of one of the four basic (essential) motifs, or introduce a new secondary design. Such innovations are not easily made, since the equilibrium of the overall design is put into question when a change in one area is introduced. The innovation is further open to the criticism of commoners, other carvers and master carvers who will evaluate the innovation according to the traditional schema. The master carver, however, is capable of introducing more profound changes than others in that he not only knows about the visual – aesthetic or iconographic – properties of a prow-board, but also has knowledge of the meanings of the motifs and their interrelation on the symbolic plane. Those are contained in the poetic formulae that he receives from his master, which metaphorically identify the design with qualities such as brightness and speed. The master carver is thus capable of interpreting the motifs of the prow-board on the metaphorical plane, and reinterpreting the visual rules that regulate the disposition of motifs (ibid.:201–2). Scoditti tells us that one of the main criteria for judging the quality of a carving is the separation or legibility of elements. One of the main innovations at the time Scoditti did his fieldwork (in the 1970s) was an element that served to separate two of the basic motifs, thus making them more legible. The imperative of legibility is also constrained by a technical requirement: if motifs are not sufficiently well carved, colours tend to run into each other, making the image look – according to the artists – confused.

An interesting parallel between Kitawan carvers and Dingshu artisans is the idea that the design of a canoe prow-board, and the ideal shape of a teapot, have been fixed in the past, or at least are said to be so, yet can also be improved upon. A master in Kitawa cannot change a design on the prow-board, he can only make clearer the overall disposition of the elements. Similarly, Gu did not change a 'traditional' design, he simply modified it to make it closer to the ideal in his understanding of the principles of *zisha* pottery. The distinction made between skill and esoteric knowledge one finds in Kitawa is not, however, replicated in Dingshu, although there is a distinction between those who 'know', in other words masters, and other common artisans. In Dingshu, theory necessarily derives from practice: artisans claimed that connoisseurs could never fully understand the craft since they did not practise it.

Another parallel that can bring us to understand Gu's ideas can be drawn with Chinese calligraphy. Yen (2005:94–5) tells us that the calligrapher must exaggerate certain elements of the written character, so that there can be no confusion about it. A line of four dots at the bottom of a character (the fire radical), for instance, must be written in a single movement, so that one can follow the line the brush made as, one by one, it made the dots. Indeed, in some cases the four dots are simply rendered as a straight line. Of course, in this case, the calligrapher is also not at liberty to change elements of the character. Instead, they must find a way of writing it in as clear a way as possible.

The master as editor

In light of these considerations, how can one conceptualize the figure of the master in Dingshu? I would like to propose for a moment the textual metaphor of editing to think about this. Contrasting 'Western' and Chinese approaches to textual criticism, Cherniak (1994) highlights the very different valuation of changes brought to classical texts. Modern Western textual criticism is geared towards recuperating the original version of a text by comparing the different versions available. The assumption is that transmission of a text is necessarily a degenerative process – Cherniak talks of corruption and contamination. The aim is recovering the author's original intentions. She contrasts this to the Chinese relative tolerance for 'collaborative authorship': it has always been understood that scholars can bring changes to texts, by correcting typographical errors, by correcting the content and clarifying the meaning if they think it is warranted. The Chinese classical author appears then in a rather different light to their Western counterpart. The idea of the text as resulting from an individual creative author who produces a text with original content gives way to the idea of the author who stabilizes and canonizes other texts: an editor. The exemplar of the Chinese author, Cherniak reminds us, is Confucius, who is known to have said, 'I transmit. I don't make new works' (ibid.:15–16). Instead of the Euro-American literary author, the Chinese author might be compared to the scientific author (Biagioli and Galison 2003).

Such a description of the work of the Chinese classical author strongly reminds me of Gu Jingzhou's approach to his craft. I think that Gu is similar to the Chinese author who does not create but stabilizes, though I would suggest there are important differences in the way this operates between the editing of texts and the making of clay objects. The creative process of editing is teleological in outlook in that it is directed by the intentions of the artist master, and can be compared with Thorstein Veblen's (1914) and Peter Betjemann's (2008) notion of workmanship in crafts. Inspired by Veblen, Betjemann (2008) argues that Euro-Americans have inherited a certain view of craft from the Arts and Crafts movement of the nineteenth century that praises irregularity and lack of full control over the work process, and values response and engagement with materials. This view of crafts contradicts another, which pre-dates the movement, that is teleological in outlook, and posits that the act of making is a process of transforming vision into form. Veblen's 'instinct of workmanship' is the instinct of getting things done and finished in the most efficient way possible. Veblen's ideas were formulated as a direct critique of central figures of the Arts and Crafts movement, whose approaches were described as an 'exaltation of the defective' (Veblen, cited in ibid.:187). According to Betjemann, a major paradox appeared in the Arts and Crafts movement:

> Reacting to the machine, the figures [including William Morris and John Ruskin] defending craft ended up placing skill in a peculiar position: ... the better and more reproducible the artisan's skill, the more mechanical – and hence less identifiably crafted – looked the result.
>
> (Betjemann 2008:184)

Gu Jingzhou's high level of skills was directed towards a certain end, the making of a 'perfect' pot where all the essential elements were well defined. Gu would tell his students that when he worked, he treated clay as if it were a child. By this he meant that he could order it about, and the clay would respond to his thoughts and intentions – 'go left', 'go right' – and the clay would obey. It was as if (again, in the quasi-magical narratives told about the master) the clay responded to Gu's intentions.

For Gu and other masters in Dingshu, creating new forms was an activity that should fit with an idea of tradition. In the process of recreating over and over classic designs in clay, master artisans explore variations, and in each embodiment of the design they select the elements that they think are appropriate and those that are not. This is a process that is dependent on the embodied knowledge of the master, who needs to be able to identify at a glance the problems of initial attempts and know how to adapt sequence and skills to bring about the desired modifications. But these variations on traditional designs are understood by artisans as being incremental, an evolution of the design towards its perfect form. As such, there is an idea that they can come to be fixed – that the end result of this evolutionary process can be judged correct and further improvements no longer possible. (I mentioned earlier the comment of Gu, that no one could improve the designs he had perfected.) Shifting perspective, one can also claim that in this case a personal innovation comes to be validated by an authority and becomes part of the canon of the aesthetic principles of the craft. For this to happen, a master needs to have the influence – the charisma – to convince others that their innovations are indeed improvements. When other artisans introduce Gu Jingzhou's changes or principles in their own work, recognizing them as 'traditional', they are, I would argue, confirming Gu's status as a master.

The master as singularity

Gu's life story explains the authoritative position he gained: one of the first masters to join the cooperative in the 1950s and one of two masters (the other being Jiang Rong) to have lived until after the Cultural Revolution to see the revival of a more traditional approach to making *zisha* ceramics, starting in the late 1970s. Gu was also one of the most important teachers during most of the Zisha Factory's existence. His authority was thus, to borrow Weber's ideal types of authority (Weber 1958), both traditional (he was one of few to have been a master before the era of the cooperative and factory) and rational-legal (because of his position of influence within the factory). But Gu's life, as narrated by the people he knew well, also hints at his charismatic authority. I am notably thinking of his practice of calligraphy, knowledge of the classics and the claims that Gu produced his greatest masterpieces after he began these practices. As Feuchtwang and Wang (2001) argue, Weber's three types of authority need not be mutually exclusive, as exemplified by the village cadres they interviewed who were often called in to solve religious or 'traditional' matters. Bourdieu (1985), with reference to Ryle, points out that social scientists tend to make a category mistake when talking of charisma. Charisma is located not in the individual but imagined to be part of the individual by the group. This explains how an individual can represent a group and can stand for something it is not:

> political fetishism lies precisely in the fact that the value of the hypostatized individual, a product of the human brain, appears as charisma, a mysterious objective property of the person, an impalpable charm, an unnameable mystery.
>
> (ibid.:740)

The master is such a hypostatized individual, their genius created by the actions of those who recognize genius within them. The master's students, through their narratives, are playing a key role in the creation of this charisma. They were telling me that through moral and cultural self-cultivation, Gu was able to achieve a perfect understanding of the principles of the craft. This charisma was expressed not only through the person, but also through the person's products. The authentication of the works of the master by his former students is crucial in constructing the figure of the master.

In the example of authentication mentioned earlier in this chapter, artisans who had been close to Gu did not simply decide on the authenticity of the pot, but pointed out certain features of the work that proved their point: these were perfection in craftsmanship, the perfect balance of proportions, a sense of harmony. In sum, these amounted to perfect adherence of the technical and aesthetic principles of the craft and a perfect rendition of them in clay. As the artisans singled out the characteristics of the pot, these were referred back to the achievements of the master and his self-cultivation through calligraphy and learning of culture and history. Through the act of authentication, the surface of the pot came to embody the achievements of the master. Hypothetically, the authenticator could be wrong: a good forgery might be passed off as authentic, and vice versa. But the object remains an agent (see Gell 1998), carrying on the charismatic authority of the master, whether authentic or not. Charismatic authority is extended not only through words and actions, but also through objects, which are the agents of the (imagined) master. The master's charisma will survive much longer in their works of fired clay – that most durable material – extending the agency of the master to convince others of their genius and authority.

Following Guyer (1993, 1996), I want to suggest that one can think of the figure of the master as a 'singularity', in the sense of the self-realization of an individual. Guyer borrows the term from Kopytoff (1986), and contrasts her own case, pre-colonial Equatorial Africa, with Melanesia. In Melanesia, it is objects that are singularized, following the accumulation of historicity through circulation which is reflected back on the Big Man. In Equatorial Africa, however, Guyer argues that objects are not singularized, because of a particular feature of these societies, whereby common items are easily transformed into items of value. For instance, a hunting spear can readily be transformed into a ceremonial spear, and vice versa. It is, however, people who become singularities, and in particular those people who are able to transform objects: 'what was singular was the skill needed to play the register' (Guyer 1993:252). Included in this group of skilled and charismatic individuals are heroes, virtuoso performers and craftsmen – notably blacksmiths. One might make a parallel here with the Confucian understanding of self-realization, which is the sublimation of self-cultivation through self-transformation, and understood as the extension of the self and communion with others and the universe (Tu 1972). In a similar way to the African blacksmith, the self-realized master in Dingshu becomes a singularity in their capacity to produce objects that become exemplars in the history of the craft.

Beyond his capacity to 'make', as I mentioned earlier, Gu was also a master at talking and writing about *zisha*. Gu developed a discourse about *zisha* that became influential. This underlines the importance accorded to 'speaking' and 'writing' in the process of becoming a master. As I came to know several of Gu's students, I was struck by the close similarities in their ways of talking about the craft. Even artisans who had not been Gu's students would repeat many of his sayings. In a way, the utterances of the master had become the 'correct' way of talking about the craft.

I have heard endless discussions about who is the true follower of Gu's legacy. Many claim to follow in the footsteps of the master, but these claims are often disputed by others. Artisans will claim of their peers that they have not properly understood Gu's teachings, or are not following them rigorously enough, and this has as a consequence in that the teachings are not passed on properly to the next generation of artisans. Yi, for instance, often complained about two members of the Yao family, father and son. The former had been one of the first apprentices of Gu, and had taught his own son, but Yi claimed that they were not properly following some of the master's teachings. In one classic design, the *shipiao*, the lid is slightly thicker than in others for aesthetics reasons, but it is hollowed out in order to make the lid lighter. Gu had taught that the way of making the lid is to combine two discs, and remove the central part of one of them. But Yi told me that he had observed the daughter and grand-daughter of the Yaos scooping out the clay from a thick disc with a tool, instead of joining two discs as taught by the master. Yi felt that this was basic knowledge among the students of Gu, and got angry that this had not been properly passed on from father to son and grand-daughter.

Mastery and the Chinese state

It was important for me to devote most of this chapter to the figure of Gu Jingzhou, both as a way to reflect on mastery – the ideal culmination of an artisan's career that starts with apprenticeship – and to think about the influence a single individual can have on the craft as a whole. Rather than accepting at face value claims of genius, I wanted to understand how genius is constructed through narratives and authentication of a master's works. This not only tells us something about the master as a person, but also about the people who are recounting narratives about them. The narratives, words and actions of the master must have meaning for other artisans, which might lead one to ask: How does the master create meaning for others? I have suggested that charismatic authority plays a key part in this process. Through charisma, the master can be a prime agent of stability in craft and other traditions, expressing judgements of value that validate certain forms as correct and traditional. I realize that by focusing on one individual I might have neglected the wider picture of what it means to be a master in Dingshu. However, a master is by definition a unique character, and this is why it was important for me to look into the life history of this person. Borrowing Jane Guyer's concept, I have found it useful to identify Gu Jingzhou as a 'singularity', a charismatic character who is in a position to redefine the craft. Gu was a singularity in the sense that he was able to transform his own interpretation of tradition, his own personal style, and his own embodied knowledge, into, in effect, a tradition.

Of course, this situation is likely to change with the next master or singularity of *zisha* craft. China is very active both in contributing to the UNESCO list of intangible cultural heritage and in creating its own national list of endangered crafts and traditional practices; as mentioned, the craft techniques of *zisha* were added to this list in 2006. The list also includes what have been called 'inheritors' (*chuanchengren*) of the craft, an honorific title attributed to figureheads of the cultural form they represent. In the case of *zisha*, a single master, Wang Yinxian, has been named as an inheritor. Wang is someone with a very different style, indeed a different approach, to Gu Jingzhou. She has specialized in the making of pots that resemble natural objects or animals, such as pumpkins, bamboo and flowers. Her story is illustrative of the ambitions of the cooperatives in the early 1950s. Her family was poor and did not have a background in pottery. When Wang joined the Zisha Cooperative at the age of 14, this was regarded as an immense privilege by the family. Her apprenticeship was cut short after the second year, at the time the Zisha Workshop became the Zisha Factory and recruited new students by the hundreds. Wang became a teacher to many students before she was able to fully develop her skills. She claims to have always given her best in learning and improving herself. She worked closely with her teacher, the master Zhu Kexin, in reproducing an antique object in the collection of the Shanghai Museum, a cup in the form of a peach, which became one of her iconic works. This and other works went on to receive prizes at international art exhibitions. These successes led her to take on leading roles in the factory, and receiving honours at national level, including the title of 'role model' for women, given by the All-China Women's Federation.

One might wonder what effects the nomination of a single master as an 'inheritor' of craft techniques will have on the future of *zisha* pottery, as the work of a specific individual is singled out by the Chinese government, and as individual creativity gets entangled with cultural nationalist policies. Without wanting to speculate on this question, it is opportune to reflect on the new relationship that has been created between the Chinese nation-state and *zisha*. Wang Yinxian, aged 63 at the time of her nomination as an 'inheritor', says she has almost completely stopped working because of poor eyesight, and is not active in instructing apprentices. Being an 'inheritor' does not appear to involve actively passing on knowledge of the craft. The webpage dedicated to her on the official website of intangible cultural heritage in China presents her most iconic works, together with photographs of her posing with Gu Jingzhou and her teacher, master Zhu Kexin.[3] The emphasis, one might note, is on products rather than intangible techniques and skills, and in fact the page

3 See www.ihchina.cn/12/13757.html (accessed 10 January 2017).

dedicated to the craft as a whole[4] presents items that are richly decorated, emphasizing the finished product rather than the process of production.

In contrast to the success of Wang Yinxian, Mr Ma, an artisan who was among the first cohort of students to join the cooperative in 1954 and had started learning the craft before that time with his parents, talks bitterly of his relative lack of success in the *zisha* art world. He claims that he might have had the success and recognition of Wang Yinxian, but the internal politics of the Zisha Factory prevented him from gaining as much fame as some of his colleagues. He cites the fact that teachers within the factory had their favourite students and had much influence in promoting their career, in particular by asking them to accompany them to national and international art fairs. Ma also cites the fact that he did not become a member of the Communist Party, a requirement for success in Dingshu. This explains why, he claims, his title of 'master' was attributed at provincial level as opposed to national level, and he is only one of a handful who figures on the provincial list of masters. He explains his relative success despite lack of recognition strictly in terms of his hard work: 'I make, make, make all day; smoke, wine, play mahjong, I don't do any of those things.' Yet many of his colleagues who joined the factory in later years sell their work for large sums of money, whilst he says he cannot even afford to buy a car. Ma refuses to tell me much more about his situation: 'there are some things I can't tell you' about the politics of the community of artisans, he said.

4 See www.ihchina.cn (accessed 10 January 2017).

CHAPTER 8

Conclusion

How does a craft reinvent itself as 'traditional' following times of heightened cultural, social and political transformations? I started the book with this question, and I can now return to consider some possible answers based on my discussion in the preceding chapters. *Zisha* is promoted as a 'traditional' craft, heir to techniques and aesthetic principles that extend back to a time when the craft was passed on from parent to child, or master to apprentice. The craft appears at first glance to be relatively conservative, and artisans complain about those who do not follow the basic principles or tradition of the craft. Yet when one looks at the profound political and social changes that have occurred within the life spans of artisans, including in terms of reshuffling and the creation of new social relationships within the community of artisans, one might realize that the conservatism of the craft is somewhat illusory.

What is promoted as a tradition is the outcome of a process, involving both the making of things with skill and the negotiation of social relations and relations of power. I wanted to explain how *zisha* has become an 'exclusive' craft, its techniques giving way to discourses about the uniqueness of craft, artisan and pot. I have found that it was important for artisans to carefully negotiate an identification with tradition at the same time as establishing their independence from it, notably with reference to the notion of *fengge* or personal style. Within that perspective, individuation is created through a dialectic between the act of creation and subsequent identification with the creation. I looked at the way in which the creation of a cooperative factory created possibilities for new social relations, the most emblematic of which were relations between classmates. When it became possible for artisans to work independently, these relations could be called upon for learning new techniques, finding suitable teachers for novices, and more generally in establishing networks of exchanges that continue to affect the way the craft is made and evaluated. In fact, these relations have an impact on the very notion of style, or at least on the way artisans think about style. Whereas before the cooperative, apprentices usually had one master and as a result developed a very similar style to that of their teacher, in the Zisha Factory they had many teachers and increased possibilities for exchanges, and there was thus room for the development of a more personal style or at least becoming proficient in different styles. However, precisely because of these new possibilities for exchange and the development of a variety of styles, further elements of distinction were needed in order to stress one's uniqueness. If distinction is no longer made exclusively on the basis of affiliation with a master's style, new criteria of differentiation are needed. Accordingly, artisans constantly argue about what constitutes 'tradition' and who is following it in an appropriate way. Artisans also negotiate exchanges with their peers, paying regular visits to some whilst refusing exchanges or dissociating themselves in discourse from them. At the same time, the artisan's style and uniqueness needs to be confirmed by members of the 'art world' – scholars, connoisseurs, collectors and, crucially, fellow artisans.

One might say that the entire career of the artisan revolves around the making of objects that are increasingly agent-like (Gell 1998) in carrying the name and fame of their makers. As seen, the career of the artisan starts with the tedious and frustrating experience of making over and over the same clay discs and strips, which are recycled at the end of the day. Artisans might even fail that first stage of apprenticeship, and be deemed not good enough to pursue learning. Gradually, the young artisan will make objects that are fired and sold; these objects will carry the name of the artisan, and together with competitions and awards, will contribute to their increasing fame. The relationship between object and person is in flux; it gradually becomes more substantial as the object becomes more 'beautiful' and carries the signs of the artisan as *fengge*, and also as the artisan becomes cultivated and able to make this cultivation apparent in pots and through discourse.

The agency of objects changes over the years. At first, it might only be the agency of a student attempting to impress their master. Later on, agency is manifested through sales, as objects come to be more valuable, through recognition by others and through identification of an object with a person. This is the case for the work of Gu Jingzhou, a central figure not only in terms of influence in the community of Dingshu, but also an agent in the stabilization of craft forms: his work, his personal style, came to define what is correct and traditional. For the master, agency appears as the capacity to make 'classic' one's personal and intimate understanding of the craft, thereby influencing future generations. The changing relationship between artisan and creation is also a changing relationship between the artisan and a collectivity. The principles of self-improvement through self-cultivation, and of the 'veneering' of the self with culture, suggest a specifically Chinese answer to the question of what is implied in the act of creation and processes of creation of agent-like objects.

The state has figured as an agent throughout this book, notably in its economic and social roles of reconfiguring systems of production and learning. In a simplistic form, the narrative of the craft's relationship with the Maoist state seems to imply a failed attempt at modernizing the craft, resulting in a return to a state of affairs that existed before 1949. Whilst the Chinese state was relatively disengaged from the production of *zisha* before the founding of the People's Republic in 1949, the Maoist years (1949–1976) are characterized by control of all aspects in the production and marketing of the craft. The economic reforms introduced at the end of the 1970s in turn might be seen as the start of a disengagement of the state from production. When the Zisha Factory stopped providing for its workers in the mid 1990s, individual artisans were once more legitimated to produce pottery in the privacy of their homes or private workshops, teaching the craft to their sons and daughters, or to apprentices chosen by them, and marketing and promoting their own products. Indeed, being an artisan in China today implies being an entrepreneur and businessperson, with clear ideas about markets and trends, and clear ideas about how to carefully manage trade relations – all aspects of production that were left to management to deal with during the years of the state-controlled factory. Yet I would argue that the involvement of the state in *zisha* has not decreased in recent years; rather, the influence of the state has extended to other aspects of the craft.

Throughout the twentieth and into the twenty-first century, a number of fundamental shifts have changed the composition and structure of the community of artisans producing *zisha* craft in Dingshu. These changes are related to changes that China as a whole underwent, including collectivization (in the 1950s), the Cultural Revolution (1966–1976), economic reforms (since 1978) and the state promotion of Chinese heritage (2000s). When I started to reflect on the situation of artisans during my initial fieldwork, I thought of the current situation in terms of a 'reappropriation' of craft by artisans following the economic reforms at the end of the 1970s. Artisans, it seemed, were reappropriating not simply the legitimacy of producing pottery, but all stages of production, transmission of knowledge (teaching one's own child, or choosing one's apprentices; see Chapter 4), and ways of talking about the craft and agency in controlling its value (Chapter 3).

Yet one might also talk of a more recent reappropriation by the state of its influence on the craft in the 2000s. The state has taken on a new role in the art world of *zisha* beyond the role of social and economic regulator it had taken in the 1950s and abandoned in the 1990s: it has become an arbiter of the cultural value of the craft, its significance for Chinese culture, and of what counts as authentic and traditional in the craft. I have identified three novel ways in which the presence of the state is felt in present-day Dingshu. Firstly, to a limited extent, the teaching of the techniques of the craft has resumed in the state education system: young people can now obtain a degree in *zisha* ceramics from the local vocational school. Titles and awards have been attributed by the county, provincial and national government since the years of the Zisha Factory, but one might also note an intensification of the influence of the state in promoting the work of individual artisans through the creation of new state-run competitions (as mentioned in Chapter 3). In these ways, the state influences what is considered appropriate, correct, beautiful and even traditional (as in the relatively new competition for hand-crafted *zisha*). Obtaining titles and obtaining awards in competitions are not of great importance for established artisans, but they are essential for the younger generations to distinguish themselves in a saturated market.

Secondly, the inclusion of *zisha* craft and its techniques on China's national list of intangible heritage is another significant way in which the state is defining what counts as authentic and traditional. This is particularly significant in the nomination of a single master of the craft as an 'inheritor'. That this particular master had achieved political influence beyond the craft community, as mentioned in the last chapter, has arguably contributed to her position of influence within it, which led to her nomination as an 'inheritor'. This gives an indication of the continuing close relationship between the politics of the craft and state politics.

Thirdly, the event I introduced this book with, the construction of a wall around Huanglongshan to protect the source of the precious *zisha* clay following accusations of foul play in a news report gives an indication as to how fast the state can reclaim elements of control it had lost or decided not to maintain in the past. In the 1950s, the first state-owned factory in Dingshu took over the extraction and distribution of clay. Control of the raw materials of the ceramics industry was a powerful way to take control of the entire production process in the township. After the dissolution of the factory system starting in the mid 1990s, the unprofitable business of extracting clay was left to private enterprise, until 2010 when mining activities in Huanglongshan became once more controlled by the township government. The significance of this renewed control over raw materials is arguably not simply economic but symbolic: the local state has become the only (legal) provider of 'genuine' *zisha* clay.

In appearance, artisans have gained renewed freedoms in making *zisha* with the return of capitalism, yet it would be more accurate to say that they have become different artisan subjects. In Chapter 1 I suggested that the ambiguity inherent in the concept of 'craft' might be useful to avoid treating skills and politics, techniques and values, separately. I put forward the ideas of the research group Matière à Penser (Julien and Rosselin 2009; Warnier 2001, 2009), and in particular their notion of 'subjectivation' through material culture, as a way of avoiding the trap of assuming that the same embodied activities are an indication of the same experiences of making. The subjectivities of artisans, in the Foucauldian sense of being a subject, have changed in the twentieth century, despite the act of making craft items remaining the same. In Dingshu, artisans have been alternatively engaged in producing handicrafts in order to make a living, or because it was expected of them by their family, or during the planned economy under Mao because they were allocated to certain factories. More recently, artisans have been more agentive, choosing craft as an activity that is more fulfilling than white-collar work in the cities, but the economic factor still plays the most important role in the contemporary economic climate in which these same difficult-to-obtain white-collar professions pay less than being an artisan, producing craft objects that are in high demand. In these contexts, craft skills are acquired within relations of power, between artisans and managers or apprentices and masters. Allocating a certain person to a specific task was, during the planned economy, based on judgements about that person's skills and abilities. Artisans who had been apprenticed in the factory recall the tedium that characterized their first year of apprenticeship when they practised over and over the basic techniques of the craft, consisting of beating the clay into perfectly flat clay sheets and discs, and at the end of the day destroying their work in order to be able to reuse the clay the next day. As entrepreneurs starting in the 1990s, artisans changed their way of being an artisan: they became fully involved in the task of seeking opportunities for learning, finding the best ways to market their work and formulate a discourse about their style or historic significance of the craft.

Anagnost (1994) argues that in post-Mao China, the class subject has become muted, replaced by other concerns including the *suzhi* of the population. This term, usually glossed as 'quality', has become central to the People's Republic of China's dynamics of governance, and reference to *suzhi* tends to justify social and political hierarchies with those of 'high' quality gaining more income, power and status than people with 'low' *suzhi*, usually peasants (Brandtstädter 2013; Kipnis 2006). In twenty-first century Dingshu, related discourses created divisions between artisans who had been educated, who knew about the history and theory of *zisha* craft and design, and those who did not. Indeed, the increasing number (and importance for artisans) of (often state-organized) competitions also stresses knowledge of the history and culture of *zisha*, and I was told that new regulations relating to the allocation of government titles (ranking artisans according to achievements) now make it difficult for artisans without a college degree to achieve the status of high-level artisan at the national level. An indicator of being a 'modern' subject in capitalist China is being cultivated; in contrast, during the Maoist years, one of the indicators of the modernity of the country was the transformation of individual handicrafts producers into a collective of industrial producers.

Another manifestation of this state of affairs is the differences between discourses on class and capitalism that Anagnost (1985, 1994) reports from her fieldwork in Dingshu in the early 1980s, and those that I heard during my own fieldwork. The artisans and managers Anagnost met talked of 'before' and 'after' the creation of the People's Republic in sharply contrasting terms (Anagnost 1994). *Jiefang qianhou* ('before liberation') was a constant in discourse. Anagnost analysed oral histories that had been collected by history students from Nanjing University in 1958, and highlights the way these recount the suffering of ordinary artisans and the inhumanity of the kiln-owning capitalists. In these accounts, the shop owners of Shanghai are also seen to expropriate the skills of artisans who were employed there (ibid.:271).

During my own fieldwork, artisans were more ambivalent about contrasts between the situation before and after 1949. It was certainly often repeated that most artisans lived in poverty before 1949, and the creation of a cooperative in the mid 1950s was talked about in overwhelmingly positive terms. But the way artisans talked about employment in the shops of Shanghai in the 1930s contrasts particularly sharply with the accounts collected by Anagnost: rather than imagining the skills of artisans being expropriated, artisans talked to me about the 1930s as the 'golden age' of *zisha* production, a time when the most talented young artisans, who would later become masters in the Zisha Factory, were given unprecedented opportunities to develop their skills and art. The internalized discourses on class that Anagnost reports have given place to discourses on heritage and the cultural value of self-development.

Glossary

bai shifu 拜师傅	paying respects to one's master
bendi 本地	locals
benshanlü 本山绿	type of clay found in the Dingshu area
bizi 篦子	bamboo rib, tool used to shape a zisha pot (lit. a comb)
ca ganjing 擦干净	wipe clean
chadao 茶道	tea ceremony
chuanchengren 传承人	inheritor
chuantong(de) 传统(的)	tradition(al)
cu 醋	coarse, unrefined
cuowu 错误	mistake
cutao 醋陶	utilitarian pottery
dagang 大缸	large multi-purpose pot
da jiti 大集体	large collective enterprise
da shentong 打身筒	technical term for patting the body of a pot into shape
dashi 大师	grand master
dong 懂	to understand
dongde ren 懂得人	connoisseurs
dui 对	correct
fanggu 仿鼓	a type of pot in the shape of a drum
fanti 繁体	traditional Chinese characters
fengge 风格	style
ganjue 感觉	feeling
getihu 个体户	self-employed
gongchang 工场	workshop
gongchang 工厂	factory
gongfu 功夫	skill, precision of movements
gongren jieji 工人阶级	working class
gongshou 工手	worker
gongye 工业	industry
guanxi 关系	personal connection
guoying 国营	state-owned enterprise
hangye 行业	industry
haokan 好看	good looking

hukou 户口	system of residence and work certificates
jia 家	house, family
jiagong 加工	to process; create semi-finished work
jian 见	to see
juche 矩车	compass cutter made of bamboo with a metal blade
kan 看	to look
kanhuo 看火	kiln master (lit. 'looks at the fire')
kandedong 看得懂	to understand
lai baixiang 来百想	socializing
laiwang 来往	social interactions (lit. come and go)
laiwanr 来玩儿	socializing
laoban 老板	boss
laodong renmin 劳动人民	labouring people
lilun 理论	theory
lishi 历史	history
longyao 龙窑	dragon-kiln (a tunnel-like kiln that is constructed on the side of a hill)
maobing 毛病	defect
mingren 名人	famous person
nipian 泥片	leather-hard clay discs
nitiao 泥条	leather-hard clay strips
nixing 泥性	lit. 'nature of the clay'
nong gao 弄高	make tall
ong (Mandarin Pinyin: weng) 瓮	pots used for the preservation of foods and wine
pang 胖	fat
paowuxian 抛物线	parabola
pengyou 朋友	friend
pusu 朴素	plain and unadorned
pi/pei 坯	greenware (unfired pottery)
qi 气	breath, energy
qiezipian 茄子片	leather-hard clay discs (lit. aubergine slices)

renqing 人情	human feelings; social relationship
renxinghua 人性化	to humanize
shaguo 沙锅	clay cooking vessel
shangpin 商品	commercial products
shen 神	spirit
shifu 师傅	master of a trade
shipiao 石瓢	a design of *zisha* pot
shougong 手工	hand work
shougongye 手工业	handicrafts (industry)
shougongyi 手工艺	handicrafts (arts and crafts)
suidao yao 隧道窑	tunnel kiln
suzhi 素质	quality
Taodu 陶都	the Capital of Pottery
taoyijia 陶艺家	ceramics artisans
tigao 提高	to improve
tuiyao 推窑	tunnel kiln
tuzi 图子	cross-section drawing
wanmei 完美	perfect
wen 文	writing; culture
wenhua 文化	culture
wuxing 悟性	power of understanding; comprehension; talent
xi 细	fine
xiandai(de) 现代(的)	modern
xing 型	form
xitao 细陶	fine pottery
xiuyang 修养	self-cultivation
xu 序	sequence
yang 养	feeding; nurture; to raise
yanjiusuo 研究所	research department
(xiao/zhong/da) yaohu (小/中/大) 窑户	(small/medium/large) kiln household
yaoye 窑业	kiln professions
yaxiaqu 压下去	press down
yijia yihu 一家一户	every household
yishu 艺术	arts and crafts
yishupin 艺术品	arts and crafts products

zhenli 真理	truth
zhuni 朱泥	type of clay found in the Dingshu area
zisha 紫砂	a type of clay (lit. 'purple sand'), or craft products that are made of this clay
zixue 自学	self-taught

References

Anagnost, A., 1985. 'Hegemony and the Improvisation of Resistance: Political Culture and Popular Practice in Contemporary China'. PhD thesis. Ann Arbor: University of Michigan.

——— 1994. Discursive boundaries and representation in post-Mao China. In J. Hay (ed.), *Boundaries in China*. London: Reaktion Books, pp. 257–79.

Anderson, B. 1983. *Imagined Communities: Reflections on the Origin and Spread of Nationalism*. London: Verso.

Anon. 2012. Shentan Zisha Gongyi Tao [In-depth discussion of *zisha* craft pottery]. Available at: blog.sina.com.cn/s/blog_7d40193201012t7r.html (accessed 1 November 2013).

Appadurai, A., 1986. Introduction: commodities and the politics of value. In A. Appadurai (ed.), *The Social Life of Things: Commodities in Cultural Perspective*. Cambridge: Cambridge University Press, pp. 3–63.

Arnheim, R. 1981. Style as a gestalt problem. *Journal of Aesthetics and Art Criticism* 39(3):281–9.

Baidu Baike. n.d.a. 紫砂门 [The Zisha scandal]. Baidu Hudong Baike [Baidu Collaborative Encyclopaedia]. Available at: baike.baidu.com/view/3660235.htm (accessed 18 September 2014).

——— n.d.b. Yang Fengnian. Baidu Hudong Baike [Baidu Collaborative Encyclopaedia]. Available at: baike.baidu.com/item/%E6%9D%A8%E5%87%A4%E5%B9%B4 (accessed 1 November 2013).

Barmé, G., 2005. Intangible, invisible, or collectible? Issues concerning intangible heritage. *China Heritage Quarterly (Online)* 2. Available at: www.chinaheritagequarterly.org/editorial.php?issue=002 (accessed 22 March 2012).

——— 2006. China's intangible cultural heritage. *China Heritage Quarterly (Online)* 7. Available at: www.chinaheritagequarterly.org/editorial.php?issue=007 (accessed 22 March 2012).

Barth, F. 1969. Introduction. In F. Barth (ed.), *Ethnic Groups and Boundaries: The Social Organization of Culture Difference*. Oslo: Universitetsforlaget, pp. 9–37.

Bartholomew, T. 1990. A concise history of Yixing Ware. In K. Lo (ed.), *The Art of the Yixing Potter: The K.S. Lo Collection, Flagstaff House Museum of Tea Ware*. Hong Kong: Hong Kong Urban Council, pp. 42–7.

Baudrillard, J., C.R. Lovitt and D. Klopsch. 1976. Toward a critique of the political economy of the sign. *SubStance* 5(15):111–16.

Becker, H.S. 1982. *Art Worlds*. Berkeley: University of California Press.

Betjemann, P. 2008. Craft and the limits of skill: handicrafts revivalism and the problem of technique. *Journal of Design History* 21(2):183–93.

Biagioli, M., and P. Galison (eds). 2003. *Scientific Authorship: Credit and Intellectual Property in Science*. New York: Routledge.

Billeter, J.F. 1989. *The Chinese Art of Writing*. New York: Skira/Rizzoli.

Bloch, M. 1991. Language, anthropology and cognitive science. *Man* 26(2):183–98.

Bourdieu, P. 1984. *Distinction: A Social Critique of the Judgement of Taste*. London: Routledge and Kegan Paul.

——— 1985. The social space and the genesis of groups. *Theory and Society* 14(6):723–44.

——— 1986. The forms of capital. In J.G. Richardson (ed.), *Handbook of Theory and Research for the Sociology of Education*. New York, NY: Greenwood Press, pp. 46–58.

Brandtstädter, S., 2004. Money plucked from the sky: shrimp farming, entrepreneurship and the circulation of know-how in a Fujian village. *Taiwan Journal of Anthropology* 2(1):41–67.

——— 2013. Counterpolitics of liberation in contemporary China: corruption, law, and popular religion. *Ethnos* 78(3):328–51.

Brossollet, M.S. 1998. Le grès pourpres de Yixing. *Revue de la céramique et du verre* 100:4–9.

Bruner, J.S., D. Wood and G. Ross. 1976. The role of tutoring in problem solving. *Journal of Child Psychology and Psychiatry* 17:89–100.

Bruun, O. 1988. 'The Reappearance of the Family as an Economic Unit: A Sample Survey of Individual Households in Workshop Production and Crafts, Cheng-Du, Sichuan Province, China'. Copenhagen: Center for East and Southeast Asian Studies, University of Copenhagen.

Carrier, D. 2003. In praise of connoisseurship. *Journal of Aesthetics and Art Criticism* 61(2):159–69.

Cavalli-Sforza, L.L., and M.W. Feldman. 1981. *Cultural Transmission and Evolution: A Quantitative Approach*. Princeton: Princeton University Press.

Cheng, H. 2010. Shangjia Menke Luoque 'Zisha Men' Shijian Ling Zisha Hangye Zaoshou Zhongchuang [Deserted businesses: zisha industries sustain heavy losses following the zisha scandal]. *Xinhuanet.com*, 2 July. Available at: news.xinhuanet.com/local/2010-07/02/c_12290958.htm (accessed 1 September 2014).

Cheng, T., and M. Selden. 1994. The origins and social consequences of China's hukou system. *China Quarterly* 139:644–68.

Cherniack, S. 1994. Book culture and textual transmission in Sung China. *Harvard Journal of Asiatic Studies* 54(1):5–125.

Chu, C. 1988. Preface. In K. Chiu, W. Ip and W. Lee (eds), *Innovations in Contemporary Yixing Pottery*. Hong Kong: Urban Council, pp. 6–9.

Clunas, C. 1991. *Superfluous Things: Material Culture and Social Status in Early Modern China*. Cambridge: Polity Press.

Cohen, M. 1992. Family management and family division in contemporary rural China. *China Quarterly* 130:357–77.

Coleman, R. 1988. *The Art of Work: An Epitaph to Skill*. London: Pluto Press.

Cooper, E. 1980. *The Wood-Carvers of Hong Kong: Craft Production in the World Capitalist Periphery*. Cambridge: Cambridge University Press.

Coy, M. 1989a. From Theory. In Coy, M. (ed.), *Apprenticeship: From Theory to Method and Back Again*. Albany: State University of New York Press, p. 1–11.

——— (ed.). 1989b. *Apprenticeship: From Theory to Method and Back Again*. Albany: State University of New York Press.

Danto, A. 1964. The Artworld. *Journal of Philosophy* 61(19):571–84.

DeGlopper, D.R. 1995. *Lukang: Commerce and Community in a Chinese City*. Albany: State University of New York Press.

Denton, K.A. 2005. Museums, memorial sites and exhibitionary culture in the People's Republic of China. *China Quarterly* 183:565–86.

Dilley, R. 2004. The visibility and invisibility of production among Senegalese craftsmen. *Journal of the Royal Anthropological Institute* 10(4):797–813.

Dingshu (Dingshu Township Gazeteer Editorial Committee) [Jiangsu Sheng Yixing Shi Dingshu Zhen Zhi Biancuan Weiyuanhui]. 1992. *Dingshu Zhen Zhi* [Dingshu township gazetteer]. Beijing: Zhongguo Shuji Chubanshe.

Dreyfus, H.L., and S.E. Dreyfus. 1986. *Mind over Machine*. New York: Free Press.

Eyferth, J. 2003. De-industrialization in the Chinese countryside: handicrafts and development in Jiajiang (Sichuan), 1935 to 1978. *China Quarterly* 173:53–73.

——— 2009. *Eating Rice from Bamboo Roots: The Social History of a Community of Handicraft Papermakers in Rural Sichuan, 1920–2000*. Cambridge, MA: Harvard University Press.

Faure, D. 1989. *The Rural Economy of Pre-Liberation China: Trade Expansion and Peasant Livelihood in Jiangsu and Guangdong, 1870 to 1937*. Oxford: Oxford University Press.

Feuchtwang, S., and M. Wang. 2001. *Grassroots Charisma: Four Local Leaders in China*. London: Routledge.

Gardner, H. 1971. The development of sensitivity to artistic styles. *Journal of Aesthetics and Art Criticism* 29(4):515–27.

Gell, A., 1992. The technology of enchantment and the enchantment of technology. In J. Coote and A. Shelton (eds), *Anthropology, Art and Aesthetics*. Oxford: Oxford University Press, pp. 40–63.

——— 1998. *Art and Agency: An Anthropological Theory*. Oxford: Clarendon Press.

Gibson, J.J. 1979. *The Ecological Approach to Visual Perception*. Boston: Houghton Mifflin.

Gillette, M. 2010. Copying, counterfeiting, and capitalism in contemporary China: Jingdezhen's porcelain industry. *Modern China* 36(4):367–403.

Gold, T. 1985. After comradeship: personal relations in China since the Cultural Revolution. *China Quarterly* 104:657–75.

Gold, T., D. Guthrie and D. Wank. 2002. *Social Connections in China: Institutions, Culture, and the Changing Nature of Guanxi*. Cambridge: Cambridge University Press.

Goody, E.N. 1982a. Introduction. In E.N. Goody (ed.), *From Craft to Industry: The Ethnography of Proto-Industrial Cloth Production*. Cambridge: Cambridge University Press, pp. 1–37.

——— 1982b. Daboya weavers: relations of production, dependence and reciprocity. In E.N. Goody (ed.), *From Craft to Industry: The Ethnography of Proto-Industrial Cloth Production*. Cambridge: Cambridge University Press, pp. 50–84.

Götz, I.L. 1981. On defining creativity. *Journal of Aesthetics and Art Criticism* 39(3):297–301.

Gowlland, G. 2009. Learning to see value: exchange and the politics of vision in a Chinese craft. *Ethnos* 74(2):229–50.

Graeber, D. 2001. *Toward an Anthropological Theory of Value: The False Coin of Our Own Dreams*. New York: Palgrave.

Grasseni, C. 2004. Skilled vision: an apprenticeship in breeding aesthetics. *Social Anthropology* 12(1):41–55.

Gu, J. 1991. Huyide xing shen qi [Pottery art form and spirit]. In J. Shi (ed.), *Zisha Chunqiu* [Annals of Zisha Ceramics]. Hong Kong: Wenhui Chunbanshe, pp. 143–6.

Guyer, J.I. 1993. Wealth in people and self-realization in Equatorial Africa. *Man* 28(2):243–65.

——— 1996. Traditions of invention in Equatorial Africa. *African Studies Review* 39(3):1–28.

Hamlish, T. 2000. Global culture, modern heritage: re-membering the Chinese imperial collections. In S. Crane (ed.), *Museums and Memory*. Stanford, CA: Stanford University Press, pp. 137–58.

Henare, A.J.M., M. Holbraad and S. Wastell (eds). 2007. *Thinking Through Things: Theorising Artefacts Ethnographically*. London: Routledge.

Herzfeld, M. 1995. It takes one to know one: collective resentment and mutual recognition among Greeks in local and global contexts. In R. Fardon (ed.), *Counterworks: Managing the Diversity of Knowledge*. London: Routledge, pp. 127–45.

——— 2004. *The Body Impolitic: Artisans and Artifice in the Global Hierarchy of Value*. Chicago: University of Chicago Press.

Hobsbawm, E. 1983. Introduction: inventing traditions. In E. Hobsbawm and T. Ranger (eds), *The Invention of Tradition*. Cambridge: Cambridge University Press, pp. 1–14.

Hsu, E. 1999. *The Transmission of Chinese Medicine*. Cambridge: Cambridge University Press.

Huang, E., 2008. 'China's China: Jingdezhen Porcelain and the Production of Art in the Nineteenth Century'. PhD thesis. San Diego: University of California.

Hutchins, E. 1995. *Cognition in the Wild*. Cambridge, MA: MIT Press.

Ingold, T. 2000. *The Perception of the Environment: Essays on Livelihood, Dwelling and Skill*. London: Routledge.

——— 2007. Materials against materiality. *Archaeological Dialogues* 14(1):1–16.

Ip, W. 1988. Development of Yixing pottery. In C. Shing, W. Ip and W. Lee (eds), *Innovations in Contemporary Yixing Pottery*. Hong Kong: Urban Council, pp. 22–43.

Jackson, M. 1983. Knowledge of the body. *Man* 18(2):327–45.

Jan, G.P. 2004. *The Chinese Commune: A Communist Experiment that Failed*. Lampeter: Edwin Mellen Press.

Jefferson, G., and T. Rawski. 1994. Enterprise reform in Chinese industry. *Journal of Economic Perspectives* 8(2):47–70.

Jing, W. 2006. Nostalgia as content creativity: cultural industries and popular sentiment. *International Journal of Cultural Studies* 9(3):359–68.

Julien, M.-P., and C. Rosselin. 2009. *Le sujet contre les objets ... tout contre: ethnographies de cultures matérielles*. Paris: Comité des travaux historiques et scientifiques.

Keane, W. 2001. Money is no object: materiality, desire, and modernity in an Indonesian society. In F.R. Myers (ed.), *The Empire of Things: Regimes of Value and Material Culture*. Santa Fe, NM: School of American Research Press, pp. 65–90.

Keller, C.M., and J.D. Keller. 1996. *Cognition and Tool Use: The Blacksmith at Work*. Cambridge: Cambridge University Press.

Kerr, R., and N. Wood. 2004. *Science and Civilisation in China, Vol. 5: Chemistry and Chemical Technology, Part 12: Ceramic Technology*. Cambridge: Cambridge University Press.

King, A.Y. 1991. Kuan-hsi and network building: a sociological interpretation. *Daedalus* 120(2):63–84.

Kipnis, A.B. 1997. *Producing Guanxi: Sentiment, Self, and Subculture in a North China Village*. Durham, NC: Duke University Press.

——— 2006. Suzhi: a keyword approach. *China Quarterly* 186(1):295–313.

Kondo, D.K. 1990. *Crafting Selves: Power, Gender, and Discourses of Identity in a Japanese Workplace*. Chicago: University of Chicago Press.

Kopytoff, I. 1986. The cultural biography of things: commoditization as process. In A. Appadurai (ed.), *The Social Life of Things: Commodities in Cultural Perspective*. Cambridge: Cambridge University Press, pp. 64–91.

Lancy, D.F. 1980. Becoming a blacksmith in Gbarngasuakwelle. *Anthropology and Education Quarterly* 11(4):266–74.

Lave, J. 1977. Cognitive consequences of traditional apprenticeship training in West Africa. *Anthropology and Education Quarterly* 8(3):177–80.

——— 1990. The culture of acquisition and the practice of understanding. In J.W. Stigler, R.A. Shweder and G.H. Herdt (eds), *Cultural Psychology: Essays on Comparative Human Development*. Cambridge: Cambridge University Press, pp. 63–82.

Lave, J., and E. Wenger. 1991. *Situated Learning: Legitimate Peripheral Participation*. Cambridge: Cambridge University Press.

Ledderose, L. 1999. *Ten Thousand Things: Module and Mass Production in Chinese Art*. Princeton: Princeton University Press.

Lévi-Strauss, C., 1966. *The Savage Mind*. Chicago: University of Chicago Press.

Li, M. 2004. Yitiao xiaoshi de ganxiang [A few minor thoughts]. *Jiangsu Taoyi Tongxun* [Ceramic art of Jiangsu Province] 2:21.

Liu, R., and S. Wu. 2000. *Yixing Zisha Wenhua Shi* [Yixing purple clay culture and history]. Hangzhou: Jiangsu Shiying Chubanshi.

Lo, K. 1988. Patronage and its influence on Yixing teapots. In C. Shing, W. Ip and W. Lee (eds), *Innovations in Contemporary Yixing Pottery*. Hong Kong: Urban Council, pp. 10–21.

Macfarlane, A., and I. Macfarlane. 2003. *Green Gold: The Empire of Tea*. London: Ebury Press.

Marchand, T., 2001. *Minaret Building and Apprenticeship in Yemen*. Richmond: Curzon.

——— 2003. A possible explanation for the lack of explanation; Or 'why the master builder can't explain what he knows': introducing informational atomism against a 'definitional' definition of concepts. In A. Bicker, J. Pottier and P. Sillitoe (eds), *Negotiating Local Knowledge: Power and Identity in Development*. London: Pluto Press, pp. 30–50.

——— 2007. Crafting knowledge: the role of 'parsing and production' in the communication of skill-based knowledge among masons. In M. Harris (ed.), *Ways of Knowing: Anthropological Approaches to Crafting Experience and Knowledge*. Oxford: Berghahn Books, pp. 181–202.

——— 2009. *The Masons of Djenné*. Bloomington: Indiana University Press.

Miller, D. 1987. *Material Culture and Mass Consumption*. Oxford: Blackwell.

Moeran, B. 1987. The art world of contemporary Japanese ceramics. *Journal of Japanese Studies* 13(1):27–50.

——— 1997. *Folk Art Potters of Japan: Beyond an Anthropology of Aesthetics.* Richmond: Curzon.

Mu, Q. 2012. Cultural protection has no borders. *China Daily*, 1 March. Available at: www.chinadaily.com.cn/cndy/2012-03/01/content_14726404.htm (accessed 22 March 2012).

Munn, N.D. 1992. *The Fame of Gawa: A Symbolic Study of Value Transformation in a Massim (Papua New Guinea) Society*. Durham, NC: Duke University Press.

Murch, W. 1995. *In the Blink of an Eye: A Perspective on Film Editing.* Hollywood, CA: Silman-James Press.

Naji, M. 2009. Gender and materiality in-the-making. *Journal of Material Culture* 14(1):47–73.

O'Connor, E. 2006. Glassblowing tools: extending the body towards practical knowledge and informing a social world. *Qualitative Sociology* 29(2):177–93.

Palsson, G. 1994. Enskilment at sea. *Man* 29(4):901–27.

Pan, C. 2003. Duiyu Yiba Shi Dabin Yuanhu de Jiedu [Deciperhing a round pot made by Shi Dabin]. *Jiangsu Taoyi Tongxun* [Jiangsu ceramics arts news] 13:24–30.

Polanyi, M. 1962. *Personal Knowledge: Towards a Post-Critical Philosophy*. London: Routledge and Kegan Paul.

Prentice, R. 2008. Knowledge, skill, and the inculcation of the anthropologist: reflections on learning to sew in the field. *Anthropology of Work Review* 29(3):54–61.

Pye, D. 1968. *The Nature and Art of Workmanship*. Cambridge: Cambridge University Press.

Riskin, C. 1971. Small industry and the Chinese model of development. *China Quarterly* 46:245–73.

Ryle, G. 1949. *The Concept of Mind.* Chicago: University of Chicago Press.

Santos, G.D. 2008. On 'same-year siblings' in rural South China. *Journal of the Royal Anthropological Institute* 14(3):535–53.

Schapiro, M. 1953. Style. In A. Kroeber (ed.), *Anthropology Today*. Chicago: University of Chicago Press, pp. 287–312.

Schran, P. 1964. Handicrafts in communist China. *China Quarterly* 17:151–73.

Scoditti, G.M.G. 1990. *Kitawa: A Linguistic and Aesthetic Analysis of Visual Art in Melanesia*. Berlin: de Gruyter.

Sennett, R. 2008. *The Craftsman*. New Haven: Yale University Press.

Shi, J. (ed.). 1991. *Zisha Chunqiu* [Annals of Zisha Ceramics]. Hong Kong: Wenhui Chunbanshe.

Singleton, J. 1989. Japanese folkcraft pottery apprenticeship: cultural patterns of an educational institution. In M. Coy (ed.), *Apprenticeship: From Theory to Method and Back Again*. Albany: State University of New York Press, pp. 13–30.

Smart, A. 1993. Gifts, bribes, and guanxi: a reconsideration of Bourdieu's social capital. *Cultural Anthropology* 8(3):388–408.

——— 1999. Expressions of interest: friendship and guanxi in Chinese societies. In S. Bell and S. Coleman (eds), *The Anthropology of Friendship*. Oxford: Berg, pp. 119–36.

Spooner, B., 1986. Weavers and dealers: the authenticity of an Oriental carpet. In A. Appadurai (ed.), *The Social Life of Things: Commodities in Cultural Perspective*. Cambridge: Cambridge University Press, pp. 195–235.

Stafford, C. 2000a. Chinese patriliny and the cycles of *yang* and *laiwang*. In J. Carsten (ed.), *Cultures of Relatedness: New Approaches to the Study of Kinship*. Cambridge: Cambridge University Press, pp. 37–54.

——— 2000b. *Separation and Reunion in Modern China*. Cambridge: Cambridge University Press.

——— 2006. Deception, corruption and the Chinese ritual economy. In K. Latham, S. Thompson and J. Klein (eds), *Consuming China: Approaches to Cultural Change in Contemporary China*. London: Routledge, pp. 42–55.

Stockman, N. 2000. *Understanding Chinese Society*. Cambridge: Polity Press.

Strathern, M. 1990. Artefacts of history: events and the interpretation of images. In J. Siikala (ed.), *Culture and History in the Pacific*. Helsinki: Transactions of the Finish Anthropological Society, pp. 25–44.

——— 1993. One-legged gender. *Visual Anthropology Review* 9(1):42–51.

——— 1996. Cutting the network. *Journal of the Royal Anthropological Institute* 2(3):517–35.

Suresh, T.G. 2002. Rediscovering nationalism in contemporary China. *China Report* 38(1):11–24.

Tang, B., C. Li and L. Ye. 1992. Yixing *zisha* chahu shiyong gongneng de yanjiu [An investigation of the practical function of *zisha* teapots]. In J. Li and X. Zhen (eds), *Gutaoci Kexue Jishu* [Ancient Ceramics Science and Technology]. Shanghai: Guoxue Jishu Wenxian Chubanshe, pp. 389–405.

Terrio, S.J. 1996. Crafting grand cru chocolates in contemporary France. *American Anthropologist* 98(1):67–79.

Toops, S. 1993. Xinjiang's handicraft industry. *Annals of Tourism Research* 20(1):88–106.

Tseng, Y. 1998. L'art du thé à Taiwan. *Revue de la céramique et du verre* 100:25–6.

Tu, W. 1972. Li as process of humanization. *Philosophy East and West* 22(2):187–201.

Varutti, M. 2014. *Museums in China: The Politics of Representation after Mao*. Woodbridge: Boydell Press.

Veblen, T. 1914. *The Instinct of Workmanship, and the State of the Industrial Arts*. New York: Macmillan.

Wacquant, L.J.D. 2006. *Body and Soul: Notebooks of an Apprentice Boxer*. Oxford: Oxford University Press.

Walder, A.G. 1984. The remaking of the Chinese working class, 1949–1981. *Modern China* 10(1):3–48.

Warnier, J.-P. 2001. A praxeological approach to subjectivation in a material world. *Journal of Material Culture* 6(1):5–24.

——— 2009. *Régner au Cameroun: le roi-pot.* Paris: Karthala.

Weber, M. 1958. The three types of legitimate rule. *Berkeley Publications in Society and Institutions* 4(1):1–11.

Weiner, A.B. 1994. Cultural difference and the density of objects. *American Ethnologist* 21(2):391–403.

Wenger, E. 1998. *Communities of Practice: Learning, Meaning, and Identity.* Cambridge: Cambridge University Press.

Wenger, E., R.A. McDermott and W. Snyder. 2002. *Cultivating Communities of Practice: A Guide to Managing Knowledge.* Boston: Harvard Business School Press.

Willerslev, R. 2007. 'To have the world at a distance': reconsidering the significance of vision for social anthropology. In C. Grasseni (ed.), *Skilled Visions: Between Apprenticeship and Standards.* New York: Berghahn Books, pp. 24–46.

Wollheim, R. 1979. Pictorial style: two views. In B. Lang (ed.), *The Concept of Style.* Philadelphia: University of Pennsylvania Press, pp. 128–45.

Xu, X., 2009. *Zisha Gongyi* [The craft techniques of zisha]. Hangzhou: Zhejiang Renmin.

Xu, X., and G. Shan. 2009. *Yixing Zisha Wu Bai Nian* [500 years of Yixing zisha]. Shanghai: Shanghai Cishu Chubanshe.

Yang, M.M. 1989. Between state and society: the construction of corporateness in a Chinese socialist factory. *Australian Journal of Chinese Affairs* 22:31–60.

——— 1994. *Gifts, Favors, and Banquets: The Art of Social Relationships in China.* Ithaca, NY: Cornell University Press.

Yan, Y. 1996. *The Flow of Gifts: Reciprocity and Social Networks in a Chinese Village.* Stanford, CA: Stanford University Press.

Yen, Y. 2005. *Calligraphy and Power in Contemporary Chinese Society.* London: Routledge Curzon.

INDEX

Page numbers in *italic* refer to illustration captions.

www.ingramcontent.com/pod-product-compliance
Lightning Source LLC
LaVergne TN
LVHW011817200926
848484LV00005B/1230
* 9 7 8 1 9 0 7 7 7 4 9 8 0 *